Print to E-Resources

Challenges and Opportunities

Editor-in-Chief

Dr. A.K. Srivastava

Editors

Dr. M. Anand Murugan

Dr. Sanjiv Saraf **Dr. Vivekanand Jain**

SHREE PUBLISHERS & DISTRIBUTORS

NEW DELHI-110 002

Based on the Proceeding on National Seminar held at Central Library Banaras Hindu University, 15-16 April, 2011

Edition : 2012

Published by :
SHREE PUBLISHERS & DISTRIBUTORS
22/4735 Prakash Deep Building,
Ansari Road, Darya Ganj,
New Delhi–110 002

ISBN : 978–81–8329–468–3

Printed by :
D.G. Printers
Delhi–110 094

Sayaji Rao Gaekwad Library
सयाजी राव गायकवाड ग्रन्थालय
Central Library
केन्द्रीय ग्रन्थालय
Banaras Hindu University
काशी हिन्दू विश्वविद्यालय
Varanasi - 221005
वाराणसी–221005
India
भारत

Dr. Ajay Kumar Srivastav
डा० अजय कुमार श्रीवास्तव
Ph.D., M.Phil (Lib. & Information Science)
Ph.D. (Sociolgy)
University Librarian
विश्वविद्यालय पुस्तकालयाध्यक्ष

Mail: librarian.bhu@gmail.com, clbhu@bhu.ac.in
Mob: 09415293976
Phone (O) (0542) - 2369433 & 2307130
Fax: (0542) 2367133

Foreword

I am very happy to receive tremendous support and appreciation from all professional colleagues and faculty members of Banaras Hindu University during the seminar on very recent topic ***"Print to E-Resources: Challenge and Opportunities"***

Library professionals from all over India participated and discussed the issues in seminar. Now we are publishing the proceedings of it in the form of book. The major topics covered under this seminar are: E-journals, E-books, E-resources, Open Access Journals, E-Learning, E-publishing, etc.

I would like to congratulate of all Library professionals who contributed papers in the proceedings and resource persons for their valuable guidance and professional cooperation.

The book is certainly useful for Faculty members of library and information science, working librarians and LIS students due to its contents on recent challenges and emerging issues in front of us. The academic and special libraries must add this book to their collection.

—Dr. A.K. Srivastav

Foreword

I am very happy to receive tremendous support and appreciation from all professional colleagues and faculty members of Banaras Hindu University during the seminar on very recent topic "*Print vs E-Resources: Challenge and Opportunities*".

Library professionals from all over India participated and discussed the issues in seminar. Now we are publishing the proceedings of it in the form of book. The major topics covered under this seminar are: E-journals, E-books, E-resources, Open Access Journals, E-Learning, E-publishing, etc.

I would like to congratulate of all Library professionals who contributed papers in the proceedings and resource persons for their valuable guidance and professional cooperation.

The book is certainly useful for Faculty members of library and information science, working librarians and LIS students, due to its contents on recent challenges and emerging issues in front of us. The academic and special libraries must add this book to their collection.

—Dr. A.K. Srivastav

Preface

It's a great opportunity to organise a national seminar at Banaras Hindu University on the emerging issues like "Print to E-resources: challenges and opportunities" for the benefit of all library professionals.

In the present seminar above 100 library professionals are participated from all corners of the country and we share the views from each other.

We are thankful to Prof. Jagtar Singh, Punjabi University, Patiala and President of IATLIS for presiding the inaugural session and main key note speaker.

We are highly obliged to all resource persons and guests : Dr. Ram Gopal Garg Head, DLIS, Jiwaji University, Gwalior; Prof Upendra Yadav & Dr. A K Sinha from TM University, Bhagalpur; Dr. G Mahesh, Scientist, NISCAIR, New Delhi; Dr. Heerak Kanti Chakravorty, Head, DLIS, SSU; Dr. A P Singh, Head, DLIS, BHU, Dr. R N Singh (Director, Physical Education, Sagar University) for their cooperation and suggestions.

We are thankful to university authorities for their financial cooperation and support.

We all enjoyed professional meeting, technical session and discussed on various issues of the real time library related problems for the improvement of the library services.

We are thankful to the all members of organising committee and team of library professionals of BHU Library system for extending their support and cooperation during the seminar.

In the last but not the least I extend my thanks to the publisher for publishing this book in precise and beautiful form.

—Editors

Contents

Organising Committee for National Seminar on Print to E-Resources : Challenges and Opportunities

1. Dr. A K Srivastava, University Librarian & Chairman
2. Dr. Sanjiv Saraf, Deputy Librarian & Coordinator
3. Dr. D K Singh, Deputy Librarian & Coordinator
4. Dr. G C Kendadamath, Deputy Librarian & Coordinator
5. Dr. M Anand Murugan Deputy Librarian & Organising Secretary
6. Dr. Vivekanand Jain, Deputy Librarian & Organising Secretary
7. Mr. Anil Agrawal, Assistant Librarian & Treasurer
8. Dr. Rajesh Singh, Assistant Librarian & Member
9. Dr. Suchita Singh, Assistant Librarian & Member
10. Mr. Jawahar Lal, Assistant Librarian & Member
11. Mr. Ram Kumar Dangi, Assistant Librarian & Member
12. Mrs. Sneha Tripathi, Assistant Librarian & Member
13. Mr. Manish Kumar Singh, Information Scientist & Member

List of Contributors

Ajay Kumar	:	Assistant Professor, Dept. of Library & Information Science, Purvanchal University, Jaunpur.
Akhilesh K.S. Yadav	:	Ph.D. Scholar, Dept. of Library & Information Science, Mizoram University, Aizawl-796 004.
Amaresh Kumar Rai	:	Assistant Librarian (Selection Grade), Rajiv Gandhi South Campus, Banaras Hindu University, Barkachha, Mirzapur.
Amitabh Gupta	:	*Semi-Professional Assistant,* Rajiv Gandhi South Campus, Banaras Hindu University, Barkachha, Mirzapur.
Amrita Mjaumdar	:	Assistant Librarian, Dr.Bhagwan Das Central Library, Mahatma Gandhi Kashi Vidyapith, Varanasi.
Anand Kumar Tripathi	:	Librarian, Army School New Cantt Allahabad.211002(U.P.).
Anil Agrawal	:	Assistant Librarian, Central Library, Banaras Hindu University, Varanasi.
Anil Kumar Singh	:	Semi-Professional Assistant, IGNOU Regional Center, B.H.U. Campus, Varanasi.
Anil Kumar Dhiman	:	Information Scientist, Gurukul Kangri University, Haridwar (Uttarakhand).
Anshu Bansal	:	Professional Assistant, Kanohar Lal Post Graduate Girls College, Meerut, CCS University, Meerut.

Arupa Majumdar	:	Assistant Librarian, Bharat Kala Bhawan, Banaras Hindu University, Varanasi.
Ashok Kumar Shukla	:	Central Library, Banaras Hindu University, Varanasi-221005.
Harish Kumar Tripathi	:	Librarian, ICAR Library, Krashi Anushandhan Bhawan, New Delhi-12.
Hans Raj	:	Information System Officer, Indian council of Agricultural Research, New Delhi.
Kumkum Agrawal	:	Librarian, Vasant Kanya Mahavidyalaya, Kamachha, Varanasi.
M. Anand Murugan	:	Deputy Librarian, Banaras Hindu University, Varanasi – 221005.
Mahesh Singh Rana	:	Deputy Librarian, IMT, Ghaziabad (Uttar Pradesh).
Manoj Kumar Dwivedi	:	Research Scholar Central Library, DDU Gorakhpur University, Gorakhpur-273009 INDIA.
Manoj Srivas	:	Cataloguer, Central Library, R. M. L. Awadh university, Faizabad.
Mohammad Nazim	:	Assistant Librarian, Law Library, Faculty of Law, Banaras Hindu University, Varanasi.
Mustaq Ahmad	:	Assistant Professor, Department of Botany, Banaras Hindu University, Varanasi.
Muzamil Mushtaq	:	Research Scholars, Department of Library and Information Science, AMU, Aligarh.
Naveen Kumar Srivastava	:	Allahabad Institute of Engineering & Technology (AIET), Allahabad.
Navin Upadhyay	:	Assistant Librarian, Institute of Technology, Banaras Hindu University, Varanasi, UP.
P. Ganesan	:	Assistant Librarian, Central Library, Alagappa University, Karaikudi–630003.

Prabhat Kumar Pandey	:	Head & Librarian, Deptt. of Library and Information Science, Govt. Gitanjali Girls PG Autonomous College, Bhopal (M.P.).
Pravin Kumar Singh	:	Internet Section, Central Library, BHU, Varanasi.
Punit Kumar Singh	:	Professional Assistant, Central Library, BHU, Varanasi-221005.
R.P. Bajpai	:	Reader in Library & Information Science, Mahtma Gandhi Chitrakoot Gramodaya V.V., Chitrakoot (Satna) M.P.
Rabindra Nath Mohanta	:	Librarian, Vasanta College for Women, Rajghat Fort, Varanasi -221001.
Rajesh Kumar Gupta	:	Librarian, Collage of Teacher Education (CTE), Uttar Pradesh, Allahabad.
Rajesh Kumar Maurya	:	Librarian, Madhu Vachspathi Institute of Engineering & Technology, Koilaha G.T.Road,Allahabad (U.P.).
Rajesh Kumar Singh	:	Assistant Librarian, Central Library, Banaras Hindu University, Varanasi.
Rakesh Kumar Mishra	:	Information Scientist Central Library, DDU Gorakhpur University, Gorakhpur-273009 INDIA.
Rakesh Pant	:	Department of Library and Information Science, Kumaun University, Nainital Uttarakhand – 263002.
Ram Kumar Dangi	:	Assistant Librarian, Central Library, Banaras Hindu University, Varanasi-221005.
Ruchika Krishna	:	Lecturer and In-charge, Deptt. of Library and Information Science, C. S. J. M. University, Kanpur.
S.N. Singh	:	Associate Professor, Dept. of Library & Information Science, Mizoram University, Aizawl-796 004.
Sanjiv Saraf	:	Deputy Librarian, Banaras Hindu University, Varanasi – 221005.

Santosh Kumar : *Professional Assistant,* Rajiv Gandhi South Campus, Banaras Hindu University, Barkachha, Mirzapur.

Saurabh Nagariya : Librarian, Shri ram college of engineering & management, banmore , Gwalior.

Sharad Kumar Sonker : Assistant Professor, Department of Library and Information Science, Babasaheb Bhimrao Ambedkar University, Lucknow-226025.

Sheikh Mohd Imran : Research Scholars, Department of Library and Information Science, AMU, Aligarh.

Sher Singh : Assistant Regional Director, IGNOU Regional Centre, Varanasi.

Shiva Parihar : Assistant Librarian, Hidayatullah National Law University, Raipur-Chhattisgarh.

Surjeet Kumar : Department of Library and Information Science, Kumaun University, Nainital Uttarakhand – 263002.

Umesh Kherajani : B.Tech-IT, Rashvarsha, Shashi Nagar Colony, Nagwan Chungi, Lanka, Varanasi.

Vijetacharya Malviya : Librarian, Devprayag Institute of Technical Studies, Phaphamau, Allahabad.

Vikas Madan : U.P. Rajarshi Tandon Open University, Allahabad .

Vineeta Jain : Assistant Librarian, College of life science: Cancer Hospital and Research Institute, Gwalior, MP.

Vinod Kumar Singh : Research Scholar, Department of Library & Information Science, University of Delhi, Delhi-110007.

Vivekanand Jain : Deputy Librarian, Banaras Hindu University, Varanasi–221005.

Yashaswi Gyanpuri : B.Tech-EC, Rashvarsha, Shashi Nagar Colony, Nagwan Chungi, Lanka, Varanasi.

1

Electronic Journals : Challenges and Opportunities

Sharad Kumar Sonker

Abstract

Electronic resources have changed the modes of collection development, management and user behaviour because of various pricing models, licensing policy and features like simultaneous access, searching, browsing, linking etc. Many libraries are subscribing and building digital collection for wider accessibility, utility and availability to fulfill diverse informational needs of the users. This paper deals with basic concepts, features, challenges, advantages and limitations of e-journals. List of major publishers/ aggregators/vendors etc. is also referred too.

Introduction

Information technology, internet, web publishing etc. have changed the paradigm from conventional to electronic printing which produces plenty of electronic publications. The documents that are in electronic format are said to be electronic resources. The term electronic resources came into usage in 1980's when first electronic journal came into existence. Now a day's, number of electronic publications are available in various files and formats.

These publications can be accessed on payment or free of cost. Due to various features electronic resources gained recognition and popularity among users within very short span of time. E-resources include e-books, e-journals, e-databases, e-thesis, dissertations and internet resources. Out of these resources, e-journals became more popular because of its flexibility and convenience. Electronic journals are available in specialized fields and have all the characteristics of print journals like cover page, contents, vol. No. issue, editors, reviews etc. They are treated as vehicle for the scientific communication and research. Most of the search engines cover the title page of the various publishers' e-journals. E-Journals can be subscribed by academic institutions, corporate and public libraries, pay-per-view etc. Some of the aggregators/service providers offer single window access for the various e-journals.

What is an Electronic Journal?

An electronic journal should perhaps meet the AACR2 definition of a serial [a publication in any medium issued in successive parts bearing numerical or chronological designations and intended to be continued indefinitely], but in practice, it is not easy to determine if a publication is an electronic journal and libraries may define the term differently. Some of the varieties of publications to be considered include:

- an electronic version of a print journal;
- a journal published only in electronic format;
- a self-published item that does not meet the standard journal definition;
- a publisher's web site for a print journal giving extracts in electronic form as 'teasers';
- coverage of full text articles in aggregator or abstracting-and-indexing databases;
- edited compilations of listserv postings;
- Articles offered on a pay-per-view basis (is this an electronic journal or document delivery?).

Type of E-journals

On the basis of accessibility, format, availability etc. electronic journals can be categorized as following:

- ***Electronic only***—complete article of any journal along with summary or abstract, available only online having no print counter part, is known as electronic journals.
- ***Electronic version of print***—when journals are provided in both forms i.e. electronic as well as print. These journals are known as electronic version of print.

Origin of E-journal

The first peer reviewed journals were *Philosophical Transactions of the Royal Society* and *Journal de Scavants,* both first published in 1665. First electronic journals, developed in the 1980s, were e-mailed to the subscribers and made available through FTP in strictly plain text format. The pioneering time for electronic journals was 1990-95, mainly dominated by non-profit making groups interested in exploiting the technology for its own sake. The commercial publishers joined in around 1996 and dominated, mainly with the direct electronic copies of their print journals. With the advent of the web along with the development of the scanning technology, adobe portable document format(PDF) and protocol, established journals first begin to appear as electronic product on the web in 1995 by John Hopkins University Press and few journals offered by OCLC's Electronic Collection Online (ECO). JSTOR journals also became publicly available that year. Since 1997, the e-journal scenario had been moving ahead with exponential growth and achieved enormous popularity and recognition as presently electronic journals are available from major publishers, on variety of subjects and at various pricing structures beneficial to the publishers as well as libraries.

Features

E-journals have various features like searching, linking, browsing, archiving, perpetual access etc. Few salient features are listed below :

Accessibility

Access to e-journals depends on internet connectivity (Bandwidth) of the institute and licensing. E-journals can be accessed through username password , IP Address or Both.

Use of E-journals

Usage of e-journals depends on licensing agreement of the library authority. Mostly the licensing policy allows access, copying, downloading, printing and interlibrary loan facility to their users.

Online E-journals

Few e-journals are available in electronic format only and do not have any print version of it. Usually, these are created in the electronic or converted from file formats to HTML, SGML, post script, XML, PDF etc. and made available in online format only.

Fast Communication

E-journals are considered as a fast mode of communication channel compared to print version as it does not require process of printing and postal services. Thus, e-journals are used for the speedy communication of outcomes of the research. Library and the users can access e-journals immediately after hosting it on the web server.

Searching

Searching is one of the prominent feature of the e-journals which facilitates simple and advanced searching of current or archived issues of e-journals which saves time of the users and helps them in quickly locating the desired information.

Browsing

Browsing is another significant feature of the e-journals which facilitates in various ways like title, author, subject, volume, issue etc. Users not familiar with searching can use browsing facility for accessing e-journals.

Linking

It is an important feature of e-journals which helps in linking internal (index, content, section, paragraph, text, etc.) as well as external web pages or digital objects.

Registration and Customization

Registration and customisation allows users to create their profile and select journals of their interest and change the interface to feel their own. After customization user can receive automatic updates of the content pages of the selected journals from the publishers by e-mail.

E-mail Alert

Registration of a specific journal allows users to receive e-mail alert of the contents updated in the journals which helps in knowledge updating.

Online Documentation

Now a days few service providers/ vendors/ aggregators offer online help and documentation facility for search interfaces and other features of e-journals.

Availability

E-journals are accessible at any time from any where. It never gets out of print or misplaced from the shelves.

Perpetual Access

E-resources offer perpetual access to subscribed journals even after discontinuation of its subscription. Journals subscribed for definite period are available for perpetual access.

Major Publishers/Aggregators/vendors

The following URLs of publishers, aggregators, venders and service providers are checked, verified and listed as on 28/03/211.

- Elsevier Science Direct—http://www.sciencedirect.com/
- Kluwer Online—http://www.kluwerlawonline.com/

- SpringerLink—http://www.springerlink.com/
- Wiley Online Library—http://onlinelibrary.wiley.com/
- JSTOR—http://www.jstor.org/
- Project Muse—http://muse.jhu.edu/
- HighWire Press—http://highwire.stanford.edu/
- Ingenta—http://www.ingentaconnect.com/
- BioMedNet—http://www.biomednet.com/
- EBSCOhost Electronic Journals Service—http://ejournals.ebsco.com/
- SwetsWise—https://www.swetswise.com/
- Ovid—http://www.ovid.com/site/index.jsp
- Electronic Collections Online [OCLC]—http://www.oclc.org/electroniccollections/

Challenges

Economic Issues : Economic issues have come up from the rise in subscription prices of print as well as electronic journals. Few publishers offer electronic journals along with the print journals with or without extra charges. One library can not fulfill diverse needs of all the users and subscribe all the journals on any particular subject. Subscription depends on pricing models and criteria like pay per view, institutional, print + online, online only consortium etc. The consortium pricing model offers many e-journals in low cost but its pricing depends on users of the concerned institutions. It can pose a big challenge if a institute has to pay as per available users without having proper infrastructure and its usage .

High Operational Costs : Electronic resources have higher operational cost compared to print resources as it includes cost of the website linking, OPAC linking, search interface, web server, bandwidth, access points (Node), specific software, management of e-resources, troubleshooting, assistance etc. Apart form these libraries have to purchase certain service forms for common search interface offered by service providers e.g. ABC service by EBESCO, JCCC by informatics etc. It enhances operational cost of the e-resources and results in increase of cost.

Acquisition and Licensing : Acquisition and licensing of the e-resources is a complex process as it includes selection, negotiation, trail, licensing, various file and format, usage statistics etc. Licensing imposes few restrictions in reference to e-resources regarding its usage, archiving, downloading, campus access, inter library loan etc. If a library fails to comply with these restrictions , the access automatically gets denied.

Searching and Linking Full Text : Searching and linking are other important features of e-resources. No single source is available for all the journals of a particular subject. Service providers offer various products which facilitate full text linking but not the searching Like web feat, JCCC, ABC list from EBESCO etc. as there are Bibliographical database which offer link to particular digital objects.

Administrative : For subscribing e-journals the administrative cooperation is very important as it has to offer support in infrastructure development and providing funds for subscription of the journals.

User Education and Support : E-resources require organisation of users' education and awareness programmes with a motive to make them familiar with electronic resources, access and legal usage etc. Support should be readily available for troubleshooting as it enhances the usage of e- resources.

User Statistics : User statistics is one of the major problem in the digital environment as few publishers, vendors, and aggregators offer user statistics in their own formats which may not be of adequate help in the comparison of e-resources. Majority of service providers do not offer user statistics and one has to depend on the publisher/aggregator/vendor, if we don't have any system to collect the usage detail of the library.

Advantages of Electronic Journals

Saves Time : E-journals are treated as fastest mode of communication. It is available on the internet, within fraction of seconds and can be uploaded on the server. The users and library need not to wait for postal services for having journals within the library and users can access desired information by searching it from anywhere and at any time. Thus, e-journals save time of the library as well as users.

Ease of Access : E-journals offer convenience, flexibility, quick Access, retrieval, downloading and printing of the relevant issue, multiple access, support of different types of search techniques and accessible from any network .

Faster Medium of Communication : Most significant advantage of E-journal is its speed of communication having all the details of print journals e.g. title page, editor, reviewers, vol. number, issue number, ISSN, etc. If the article is uploaded on the publishers' server it can be read, commented on by the readers and amended at the same time.

Saves Storage Space : Space saving is another significant advantage compared to space required for the print journals , as in case of e-journals you can store lots of articles in electronic format.

Attractive Presentation : Electronic journals include various types of digital objects (text, audio, video) which make it attractive and impressive to the human mind.

Non Panic : It is of vital importance that librarian needs not to worry about damage, postal delay, misplacing or theft etc.

Limitation of Electronic Journals

1. Inadequate Infrastructure, journal titles and back files.
2. Users are more comfortable with print than computer monitor.
3. Coverage is incomplete in comparison to print version.
4. Authentication of the issues is a big problem.
5. Long term preservation is not sure.
6. Information literacy is another major problem.

7. Complicacy of subscription and licensing of the e-journal.
8. Scrolling is another problem, as computer cannot display complete pages.

Conclusion

E-journals have certain advantages over the print version. Number of E-journals is available free or subscription base which can be accessed from anywhere any time. These journals may also be clubbed together in a single device. We spend plenty of time in going through print-journals while e-journals are highly convenient. E-journals are being used as a vehicle of scientific communication thereby sifting the paradigm of publications.

References

1. **Curtis, Donnelyn, Virginia M. Scheschy and A. R. Tarango.** *Introducing Electronic Journals*. Developing and Managing Electronic Journal Collections. Neal-Schuman, 2000. p1-18.
2. **Curtis, Donnelyn.** *E-journals: A How-to-do-it Manual for Building, Managing, and Supporting Electronic Journal Collections*. London: facet publishing (2005):1-36.
3. **Hitchcock, S., Leslie Carr and Wendy Hall.** Web Journals Publishing; A UK Prospective. Serials, 10.3(1997): 285-299.
4. **Jewell, Timothy D.** *Selection and Presentation of Commercially Available Electronic Resources: Issues and Practices. Digital Library Federation and Council on Library and Information Resources*. July 2001. 55p. Accessed on 30/3/2011. http://www.clir.org/pubs/reports/pub99/contents.html
5. **Paul Metz and Paul M. Gherman,** 1991. *Serial Pricing and the Role of the Electronic Journal*, College and Research Libraries, vol. pp. 315-327.
6. **Schauder, D.** Electronic Publishing of Professional Articles: *Attitudes of Academic and Implication for the Scholarly Communication Industry*. Journal of the American society for Information science, 42.2(1994):73-100.
7. **Wells, Alison.** *Exploring the Development of the Independent*, Electronic, Scholarly Journal, Electronic Dissertations Library. Accessed on 30/3/2011. http://panizzi.shef.ac.uk/elecdiss/edl0001/ch0402.html.

2

Emergence of Social Software and Their Application in LICs for Embellishing Services

Ajay Kumar, S.N. Singh &
Akhilesh Kumar Singh Yadav

Abstract

Libraries and information centres are experiencing a revolution in the way information is delivered, accessed, obtained and stored. Web 2.0, social media, cloud computing, the mobile web and new formats like ebooks all influence managerial decisions about the most effective plan for staff skills, budgeting and marketing. With the emergence of new media tools such as Twitter, Facebook, Blogs, Wikis, Flickr and RSS feeds, it is now necessary to offer a more customer-driven, socially rich and collaborative model of service and content delivery in online and in our buildings. In order to stay relevant and respond accordingly to the changing needs of users, libraries and corporate information centres need to choose the right tools from the vast array of options available and seek practical methods to implement these processes within their organisations.

Keywords: *Web 2.0, Library 2.0, Social Software, Social Networking*

Introduction

The Internet, and its most common manifestation in the form of the World Wide Web, has made a profound contribution to modern life. Increases in web usage continue to be dramatic with, today, more than 1966 million* users of a tool that has really only existed for a decade. Those of us who have seen the Internet evolve often find it difficult to remember life before, and those who have grown up with it accept it and integrate it into their lives without question. The trends we observe in the ways that people today communicate, interact, acquire and share knowledge, search, investigate, and participate in the creation and re-mixing of new content, will clearly have an impact upon the library, its services, and expectations of both.

Emergence of Social Software

The emergence of social software is a fruitful area for library professionals and Library 2.0. Although an exact definition of social software seems to be missing from the literature, the basic idea is that individuals jointly contribute to an effort, such as a wiki. The end product is not the output of any single individual, but rather is a community effort. Interestingly, in some cases, no end product exists; the product is under ongoing evolution as more and more individuals add their contributions.

In the absence of a widely accepted definition of social software, it is useful to examine common characteristics of such applications. Social computing has a decentralized organization, a transient membership and a loosely defined structure. Further, the scope of the system is rather fluid (Parameswaran and Whinston, 2007a). Content is dynamic because "user generated contents" and quality control is unstructured, occurring primarily through peer feedback. Social computing applications/platforms are quite varied. Advancement in social software caused of Library 2.0 via application of Web 2.0 in library and information centre.

Web 2.0

In Web 2.0, the Web becomes the center of a new digital lifestyle that changes our culture and touches every aspect of our lives. The Web moves from simply being sites and search engines to a shared network space that drives work, research, education, entertainment and social activities-essentially everything people do. You and your mobile and non-mobile devices—PDA, MP3, laptop, cell phone, camera, PC, TV, etc.—are always online, connected to one another and to the Web.

Web 2.0 is the label attached to these new capabilities of the next generation World Wide Web (Miller, 2006). The October 2005, O'Reilly summarises his paper about Web 2.0; "Web 2.0 is the network as platform, spanning all connected devices; Web 2.0 applications are those that make the most of the intrinsic advantages of that platform: delivering software as a continually-updated service that gets better the more people use it, consuming and remixing data from multiple sources, including individual users, while providing their own data and services in a form that allows remixing by others, creating network effects through an "architecture of participation", and going beyond the page metaphor of Web 1.0 to deliver rich user experiences (O'Reilly, 2005)." Web 2.0 generally refers to a second generation of services available on the World Wide Web that let people collaborate, and share information online.[2] Web 2.0 is the a method by which data and services previously locked into individual web pages for reading by the human beings can be liberated and then reused.

Library 2.0

Library is a social institution charged with the responsibility to serve its patrons with books and kindred material so well that they become its regular customers. The trinity of books, readers and staff collectively constitute the library. A library exists only at the moment its services are introduced to a purposeful seeker by a helpful librarian.

Library 2.0 first used by Michael Casey (2005), "Beyond websites, beyond even the world of technology, the concept of Library 2.0 embraces something...disruptive technology." Casey further stated that disruptive technologies rock the boat; they create new expectations and new boundaries. Disruptive technologies allow the customers, the user, to see beyond the limits of the old framework. A disruptive technology following the Web 2.0 concept may allow that customer to move beyond the limited role of "user", and move into the world of designer or moderator.

Web 2.0 technologies such as– RSS, wikis, blogging, personalisation, podcasting, streaming media, ratings, alerts, folksonomies, tagging, social networking software and the rest – could be useful in an enterprise, institutional research or community environment and could be driven or introduced by the library (Abram, 2007). Therefore, Library 2.0 means the incorporation of blogs, wikis, instant messaging, RSS and social networking into library services (Notess, 2006). Library 2.0 facilitates and encourages a culture of participation, drawing upon the perspectives and contributions of library staff, technology partners and the wider community. Blogs, wikis and RSS are often held up as exemplary manifestations of Web 2.0. A reader of a blog or a wikis is provided with tools to add a comment or even, in the case of the wikis, to edit the content. This is what we call the Read/ Write Web. Library 2.0 is about encouraging and enabling a library's community of users to participate, contributing their own views on resources they have used and new ones to which they might wish access.

Library 2.0 simply means making your library's space (physical and virtual) more interactive, collaborative, and driven by community needs. The basic drive is to get people back into the library by making the library relevant to what they want and need in their daily live to make the library a destination and not an afterthought.

Difference Between Library 1.0 and Library 2.0

Library 2.0 seeks to break down barriers: barriers librarians have placed on services, barriers of place and time, and barriers inherent in what we do. In this user-centered paradigm, libraries can get information, entertainment and knowledge into the hands of users wherever they are by whatever means works best (Tajer, 2009). Although Library 2.0 utilizes Web 2.0 technologies, it is not about replacing the traditional technology adapted by libraries already in use but rather about adding additional functionality. In fact, web 2.0 principles offer libraries many opportunities to better serve their existing audiences. Libraries should be seizing every opportunity to challenge these perceptions, and to push their genuinely valuable content, services and expertise out to places where people might stand to benefit from them; places where a user would rarely consider drawing upon a library for support (Miller, 2005).

Table 1 : Difference Between Library 1.0 and Library 2.0

Library **1.0**	*Library* **2.0**
Closed stacks	Open stacks
Collection development	Library suggestion box
Collection develop through subscription, gratis	User generated contents
Pre-organized ILS (Integrated Library Services)	User tagging
Walk in Service	Globally available service
Read only catalogue	Amazon style comments
Print newsletter mailed out	Team-build blog
Easy = dumb users	Easy = smart systems
Limited service options	Broad range of options
Information as commodity	Information as conversation
Monolithic applications	Flexible, adaptive modules
Mission focus is output	Mission focus is outcome
Focus on approach user	Focus on finding the user
ILS is core operation	User services are core

Source : *Soundararajan and Somasekharan, 2006.*

Application of Social Software in Library

There are many applications of social software in social networking or in social interaction as open source, blogs, wikis,

RSS, flickr, collaborative favorites, tagging, instant messages, social networks etc. These applications may be directed at both personal and work-related uses for example, it may be that social networking systems are an important new element in knowledge management systems (Smith and McKeen, 2007). Blogs (Web logs), which are basically public online journals are an example. Typically, authors allow others to comment on blog entries, which increase the social nature of the platform.

Social Networking

Professional and social networking sites that facilitate meeting people, finding like minds, sharing content—uses ideas from harnessing the power of the crowd, network effect and individual production/user generated content. Social networking could enable librarians and patrons not only to interact, but to share and change resources dynamically in an electronic medium. Users can create accounts with the library network, see what other users have in common to their information needs, recommend resources to one another, and the network recommends resources to users, based on similar profiles, demographics, previously-accessed sources, and a host of data that users provide (Manes, 2006).

RSS

RSS is a family of formats that allow users to find out about updates to the content of RSS-enabled websites, blogs or podcasts without actually having to go and visit the site. Instead, information from the website is collected within a feed (which uses the RSS format) and 'piped' to the user in a process known as syndication. Web feeds allow software programs to check for updates published on a website. To provide a web feed, a site owner may use specialized software (such as a content management system) that publishes a list (or feed) of recent articles or content in a standardised, machine-readable format. The feed can then be downloaded by websites that syndicate content from the feed, or by feed reader programs that allow Internet users to subscribe to feeds and view their content.

Wikipedia defined about RSS that it's a feed which contains entries that may be headlines, full-text articles, excerpts, summaries, and links to content on a website, along with various metadata. The Atom format was developed as an alternative to RSS.

Wikis

Wikis are another example of social computing. A wiki is a webpage or set of webpages that can be easily edited by anyone who is allowed access. Wikipedia's popular success has meant that the concept of the wiki, as a collaborative tool that facilitates the production of a group work, is widely understood. Wiki pages have an edit button displayed on the screen and the user can click on this to access an easy-to-use online editing tool to change or even delete the contents of the page in question. Simple, hypertext-style linking between pages is used to create a navigable set of pages.

A wikis is an online information compendium, such as Wikipedia, which is an online, open source, user-created encyclopedia. Social bookmarking services, which enable different users to share their Web bookmarks, are another example, as are YouTube, Flickr, MySpace, and the like (Parameswaran and Whinston, 2007b). This is a powerful tool for scholarly communication in the academic environment. Staff very positive about wikis in library Reference librarians can approach their knowledge base in a Wikipedia-like manner where the reference questions, for example, serve as starting point for a collaboratively developed knowledge base.

Blogs

A blog is a type of website, usually maintained by an individual with regular entries of commentary, descriptions of events, or other material such as graphics or video. Entries are commonly displayed in reverse-chronological order. The term web-log or blog is also refers to a simple webpage consisting of brief paragraphs of opinion, information, personal diary entries, or links, called posts, arranged chronologically with the most recent first, in the style of an online journal. Most blogs also allow visitors to add a comment below a blog entry. This posting and commenting process contributes to the nature of blogging (as an exchange of views). It is also called as 'weighted conversation' between a primary author and a group of secondary comment contributors, who communicate to an unlimited number of readers. Valuable tool for getting clients to engage with reference staff already many general blogs in library Reference librarians could set up subject-specific blogs advocating their use for scholarly discussions and commenting on research findings

Instant Messaging (IM)

Instant Messaging (IM) or synchronous massaging allows real time conversation between individuals via Internet. In addition, IM includes file transfer and the capability for video chat or voice chat. Instant Messaging allows reference services in an online media to closely approximate the more traditional services of the physical library. The time will almost certainly soon come when Web reference is nearly indistinguishable from face-to-face reference; librarians and patrons will see and hear each other, and will share screens and files (Maness, 2006). In addition, the transcripts these sessions already provide will serve library science in ways that face-to-face reference never did.

Streaming Media

Streaming media are multimedia that are constantly received by and normally provided to an end-user while being delivered by a streaming provider. The name refers to the delivery method of the medium rather than to the medium itself. The distinction is usually applied to media that are distributed over telecommunications networks, as most other delivery systems are either inherently streaming (e.g. radio and television) or inherently non-streaming (e.g., books). The verb to stream is also derived from this term, meaning to deliver media in this manner. Internet television is a commonly streamed media.

Tagging and Folksonomy

A tag is a keyword that is added to a digital object (e.g. a website, picture or video clip) to describe it, but not as part of a formal classification system. Users tag documents, choosing and adding uncontrolled keywords that allow them better to identify the documents from their own point of view. Social bookmarking systems share a number of common features. They allow users to create lists of 'bookmarks' or 'favorites' to store these centrally on a remote service (rather than within the client browser) and to share them with other users of the system for the social aspect. These bookmarks can also be tagged with keywords, and an important difference from the 'folder'-based categorisation used in traditional, browser-based bookmark lists is that a bookmark can belong in more than one category. Tagging also enables user to identify reference resources tagged by others. Therefore, users recommend reference resources to one another. It is clear that reference librarians can better explore the users' information needs and favorites.

Mashups

Web services that pull together data from different sources to create new services (i.e. aggregation and recombination). In web development, a mashup is a web page or application that combines data or functionality from two or more external sources to create a new service. The term mashup implies easy, fast integration, frequently using data sources to produce results that were not the original reason for producing the raw source data. Mashups and portals are both content aggregation technologies. Portals are an older technology designed as an extension to traditional dynamic Web applications, in which the process of converting data content into marked-up Web pages is split into two phases: generation of markup fragments and aggregation of the fragments into pages.

Podcasts

A podcast is a pre-recorded audio program that's posted to a website and is made available for download so people can listen to them on personal computers or mobile devices. What distinguishes a podcast from other types of audio products on the internet is that a podcaster can solicit subscriptions from listeners, so that when new podcasts are released, they can automatically be delivered, or fed, to a subscriber's computer or mobile device. Usually, the podcast features an audio show with new episodes that are fed to your computer either sporadically or at planned intervals, such as daily or weekly.

Advantage and Disadvantage of Library 2.0

While proponents of Web 2.0 heavily advertise the advantages, there are also those who feel that this technology will do more harm than good (Miranda and Gualtieri, 2008).

Table 2 : Advantage and Disadvantage of Library 2.0

	Advantages	*Disadvantages*
LIS Professionals	Collaboration	Too many different tools
	Customization	Doubts over the reliability of tools
	Communication	Difficulties in standardization
	Knowledge	Low level of security and privacy
	Sharing	Low level of cataloguing information
	Updating	Doubts over the longevity of tools
	Flexible Tools	Confidentiality concerns
	Speed	Ownership of data issues
	Reduction of Costs	
	Training	
	Facility experimentation	
Users	Requires little technical expertise	Rumours
	Flexibility	Security and legal issues
	User involvement	Dependence
	Time saving	Second-hand information
	Reduces information overload	

Web 2.0 is a powerful resource that will allow our customers to receive information from many sources, to be actively involved in creating content and generating knowledge, and to communicate with each other and spread ideas. These advantages bring with them certain risks, such as low quality of information, loss of data, security and legal issues.

Conclusion

The library sector needs to challenge the current presumptions about libraries, sweeping aside those that no longer make sense and determining if and how it makes sense to work around those that remain. That does not mean we will have to revise all aspects of library services, but we must now be prepared, collectively, to adapt to the changes in expectation and usage of information resources and technological changes sweeping our environment. Library 2.0 going to become commonplace so libraries, technology providers, policy makers and information consumers must work together. We must all combine to debate the good and the bad in new innovations and the possibilities they lay before us. We need to collaborate to lower the barriers that prevent others from easily being able to build upon past work.

References

1. **Abram, S.** (2007). Web 2.0, Library 2.0 and Librarian 2.0: Preparing for the 2.0 World. Retrieved September 6, 2010, from http://www.online-information.co.uk/online09/files/freedownloads.new_link1.1080622103251.pdf
2. **Casey, M.** (2005). Service for the Next Generation Libraries: Libraries 2.0 Perspective. Retrieved August 20, 2010, from http://www.librarycrunch.com/2005/10/working_towards_a_definition_o.html
3. **Maness, J. M.** (2006). Library 2.0 Theory : Web 2.0 and its Implications for Libraries. Webology, 3(2). Articles 25, Retrieved October 22, 2010, from http://www.webology.ir/2006/v3n2/a25.html

4. **Miller, P.** (2005). Web 2.0: Building the New Library. *Ariadne,* 45. Retrieved September 8, 2010, from http://www.ariadne.ac.uk/issue45/miller/

5. **Miller, P.** (2006). Coming Together Around Library 2.0: A Focus for Discussion and Call to Arms. *D-Lib Magazine, 12(4).* doi:10.1045/april2006-miller.

6. **Miranda, G. F., & Gualtieri, F.** (2008). The Revolution of the Web 2.0 in the Library and Information Services. Retrieved September 8, 2010, from http://www.terkko.helsinki.fi/bmf/EAHILpapers/Giovanna_F_Miranda_paper.pdf

7. **Notess, G. R.** (2006). The Terrible Twos: Web 2.0, Library 2.0, and more. *Online*, *30*(3). Retrieved September 3, 2010, from http://www.infotoday.com/Online/may06/OnTheNet.shtml

8. **O'Reilly, T.** (2005). Web 2.0: Compact Definition. O'Reilly Radar blog. Retrieved August16, 2010, from http://radar.oreilly.com/archives/2005/10/web_20_compact_definition.html

9. **Parameswaran, M., & Whinston, A. B.** (2007a). Social Computing : An Overview. *Communications of the Association for Information Systems*, *19*(37), 762-780.

10. **Parameswaran, M., & Whinston, A. B.** (2007b). Research Issues in Social Computing. *Journal of the Association for Information Systems*, *8*(6), 336-350.

11. **Smith, H., & McKeen, J.** (2007). Developments in Practice XXVI: Social Networks : Knowledge Management's "Killer App". *Communications of the Association for Information Systems*, *19*(37), 611-621.

12. **Soundararajan, E., & Somasekharan, M.** (2006). Library 2.0: Myth or Reality? Retrieved October 3, 2010, from http://library.igcar.gov.in/readit2007/conpro/s4/S4_1.pdf

13. **Tajer, P.** (2009). Reference Services 2.0: A Proposal Model for Reference Services in Library 2.0. Retrieved July18, 2010, from http://www.inflibnet.ac.in/caliber2009/CaliberPDF/38.pdf

14. *Figures at http://www.internetworldstats.com/ Retrieved from October 3, 2010 estimate that there were 1,966,514,816 Internet users worldwide till June 30, 2010.

3

E-Resources and Their Search Techniques

Vivekanand Jain

Abstract

Information sources are found in various formats from print to e-resources and accordingly their retrieval system exists. For library professionals search techniques are very important for the effective and precise information retrieval. Due to impact of Information Communication Technology the form of resources, indexing system and their retrieval techniques are also changing. In the present paper we discussed few important information resources with their specialized rctrieval system.

Introduction

Information retrieval is a field concerned with the structure, analysis, organization, storage, searching and retrieval of information. Earlier information retrieval meant text search only, but now it involves multimedia documents, significant text content, and other media also.

Banaras Hindu University Library is very rich in its resources. It includes handwritten manuscripts, early printed resources, modern printed resources and online information resources. BHU

Library is a part of UGC-INFONET, INDEST and ERMED Consortia for e-journals subscription. Besides these resources BHU subscribes publications from various publishers and societies on the basis of users need. The access facilities are available to BHU users from Campus wide network. It is an IP address based service.

E-Resources includes :-

1. Full-text Journals (from various publishers and Societies)
2. Bibliographic database (CAB, MathSciNet, Scifinder)
3. Citation database (Web of Science, Scopus, Google Scholar)
4. E-Books (Sage, Cambridge Univ. Press)
5. E- Thesis (ProQuest Dissertations & Theses)
6. Publishers Database (Elsevier, Sage, Springer, OUP,)and
7. Resources on Internet.

E-resources are available online as well as offline in CDROM database.

Information Retrieval System

E-resources are having special feature of fast search as well as hyperlink to its related resources. One can get information from current issue as well as back volumes in a single click. Some database/resources have its own retrieval system. The search techniques and the corresponding search operators depend on the search system chosen. The following are the basic search techniques applied in online search as well as CD ROM database search for information retrieval.

Simple Keyword Search

A search can be conducted by entering a single search term or a phrase comprising more than one term. The keyword search is the simplest form of search facility offered by most of the search engines/database.

In the following example (diagram-1) one can see the simple search in CAB database.

Diagram 1 : Simple Search

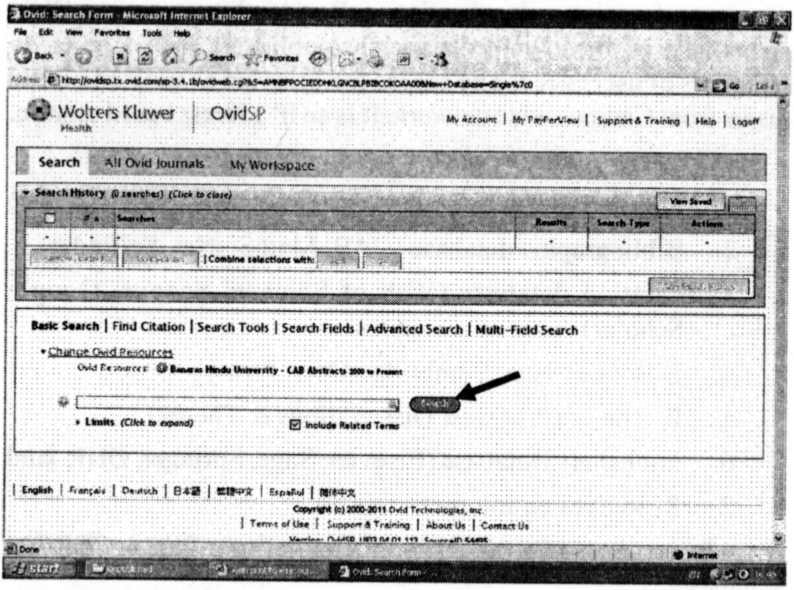

Advanced Search

In the advanced search for specific information retrieval, generally users apply the Boolean terms (AND, OR, NOT) or mathematical symbol (+, —, ^) with search terms to get the desired precise results.

In the following example (diagram-2) one can see in the advanced search in INDCAT database. Users can apply various approaches in search terms and fields. If one wants thesis on "Agam Sahitya" in the field of Title; and "Sharma" in the field of researcher; and "Varanasi" in the name of University. With the help of advanced search one can get the precise result without wasting of time.

In the advanced search, the search terms can be connected via AND / OR / NOT as per users information need. In other database/ search engines various other facilities are available to

Diagram 2 : Advanced Search

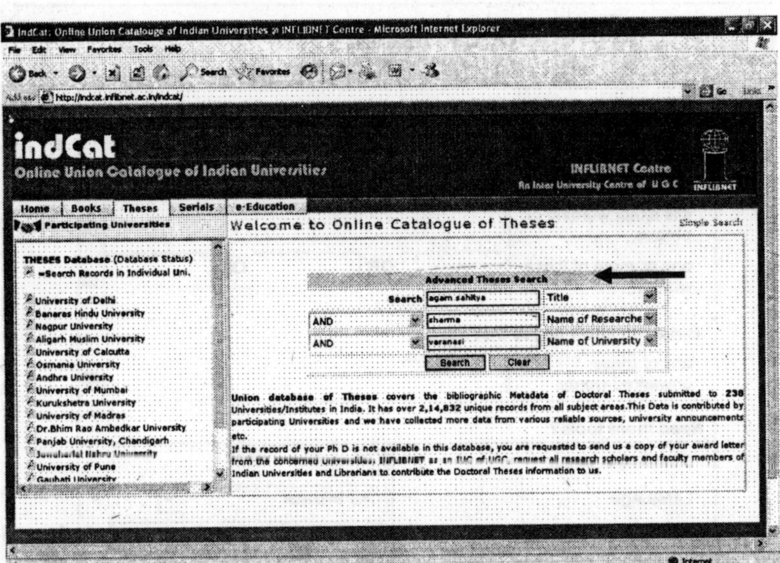

limit the search results. In the following example of "Advanced search from Google search engine" (diagram 3) in which various options are visible like date range, domain name, language, file type, region, numeric range etc.

Proximity Operators

Some search engines/ database support following search features besides above search. It may include:

- ***Followed by*** : One of the terms must be directly followed by the other.
- ***Near*** : One of the terms must be within a specified number of words of the other.

Phrase Search

Most of the search engines allow searchers to enter a phrase within double quotes in the simple or advanced search interface.

Diagram 3 : Advanced Search in Google Search Engine

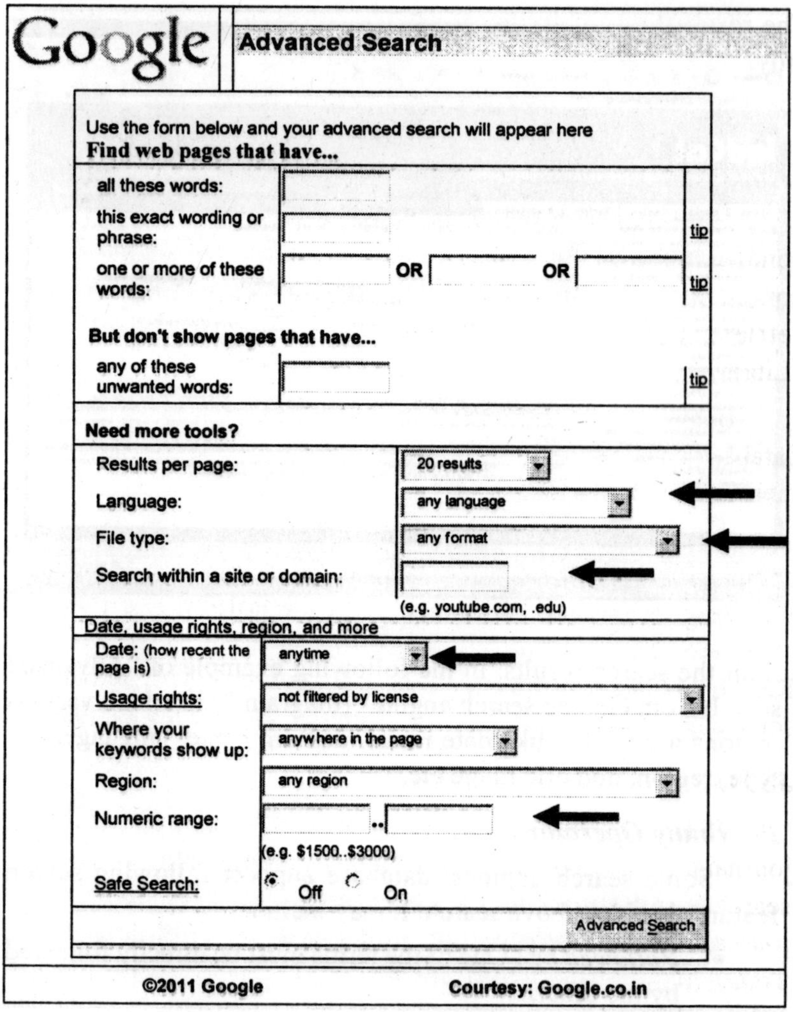

Google **Advanced Search**

Use the form below and your advanced search will appear here

Find web pages that have...

all these words:

this exact wording or phrase: tip

one or more of these words: OR OR tip

But don't show pages that have...

any of these unwanted words: tip

Need more tools?

Results per page: 20 results

Language: any language

File type: any format

Search within a site or domain: (e.g. youtube.com, .edu)

Date, usage rights, region, and more

Date: (how recent the page is) anytime

Usage rights: not filtered by license

Where your keywords show up: anywhere in the page

Region: any region

Numeric range: .. (e.g. $1500..$3000)

Safe Search: Off On

Advanced Search

©2011 Google Courtesy: Google.co.in

With the help of "phrase search" users can get much better results. The result shows only those items which contains exact phrase only.

"Academic Libraries in India"

Truncation

Truncation is a search facility that enables a search to be conducted for all the different forms of a word having the same common root. As an example, the truncated word 'Librar*' will retrieve items containing the terms 'Library', 'Libraries', 'Librarian', etc.

Besides above search (simple and advanced), some specific database designed their own search system. We describe few of them here as a special case.

Chemical Abstract/ SciFinder Scholar is a computer-based client-server interface to the world's largest database of chemistry literature, Chemical Abstracts. Over 19 million citations to chemistry publications are searchable by topic, author, CAS Registry Number, patent number, and CAS abstract number. In some cases, links to the full-text of journal articles are available.

JSTOR

JSTOR (Journal Storage) is an online database of academic journals. It provides access to its member institutions full-text searches of digitized back issues of several hundred well-known journals, dating back to 1665 in the case of the Philosophical Transactions of the Royal Society. JSTOR is presently run by independent, self-sustaining not-for-profit organization to help academic and research community. JSTOR is a fully searchable from general search engine like Google. The database contained above 1,200 journal titles in which search and browse facility available.

Diagram 4 : JSTOR Search

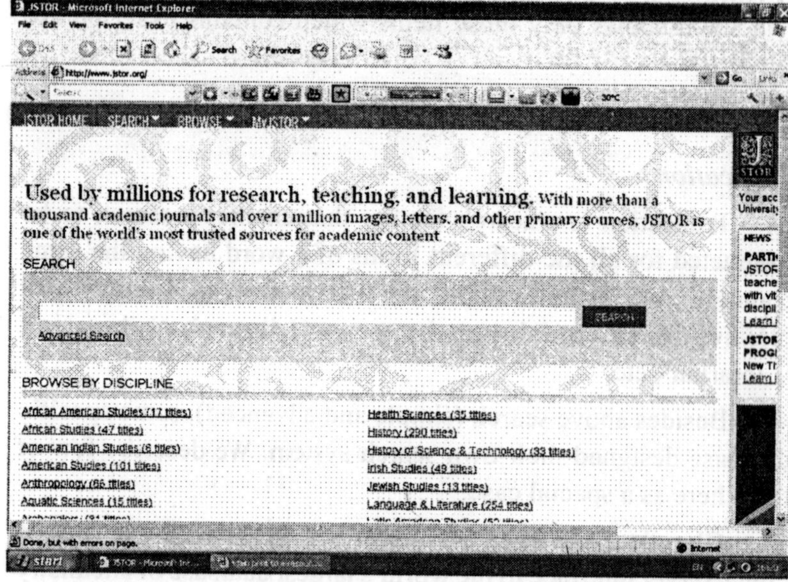

Diagram 5 : Proquest Dissertations and Thesis Database

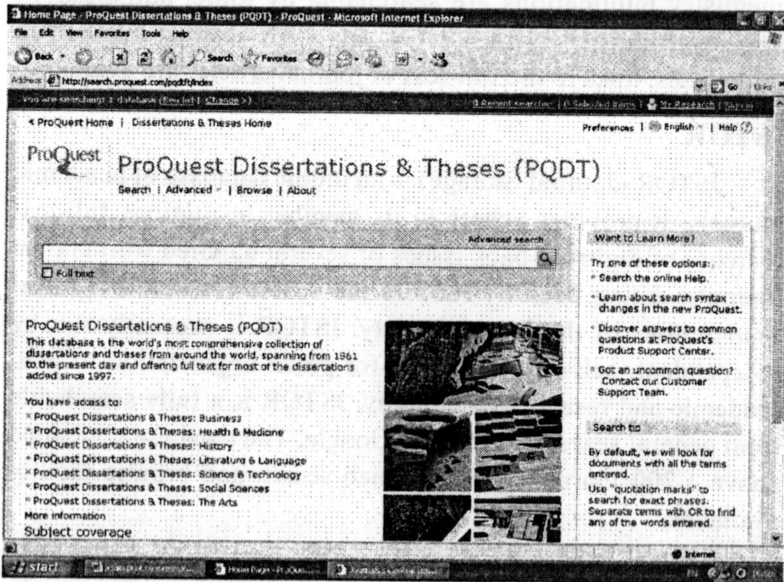

Proquest Thesis Database

Proquest thesis and dissertation database contains organized collection of research material. It includes millions of searchable citations to dissertation and theses from around the world from 1861 to the present day. It consists over 1.2 million dissertations in full text that can be downloaded in PDF format. It includes works by authors from more than 1,700 graduate schools and universities of the world.

Web of Science

It is a citation database covering above 9000 highly reputed journals of all discipline from all over the world. It provides comprehensive information regarding publication of an author and their citations. With the help of this database, one can analyze the publications of an institution, h-index etc. It also publishes Journal Citation Report (JCR) with impact factor.

Diagram 6 : Simple Search in Web of Knowledge

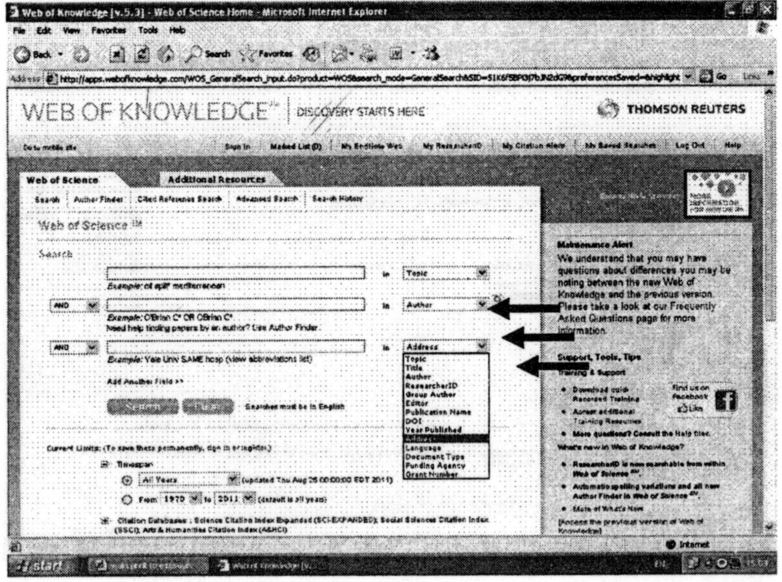

In the simple search one can apply search term from author name, affiliation, subject keyword, etc. to get the appropriate results. The search results can be sort as per user need: for example publication date, times cited, relevance etc.

In the advanced search following options are added to get the specified results.

Booleans : AND, OR, NOT, SAME, NEAR	
Field Tags	
TS=**Topic**	SG= **Suborganization**
TI= **Title**	SA= **Street Address**
AU= **Author**	CI= **City**
RID= **Researcher ID**	PS= **Province/State**
GP= **Group Author**	CU= **Country**
ED= **Editor**	ZP= **Zip/Postal Code**
SO= **Publication Name**	FO= **Funding Agency**
DO= **DOI**	FG= **Grant Number**
PY= **Year Published**	FT= **Funding Text**
AD= **Address**	OG= **Organization**

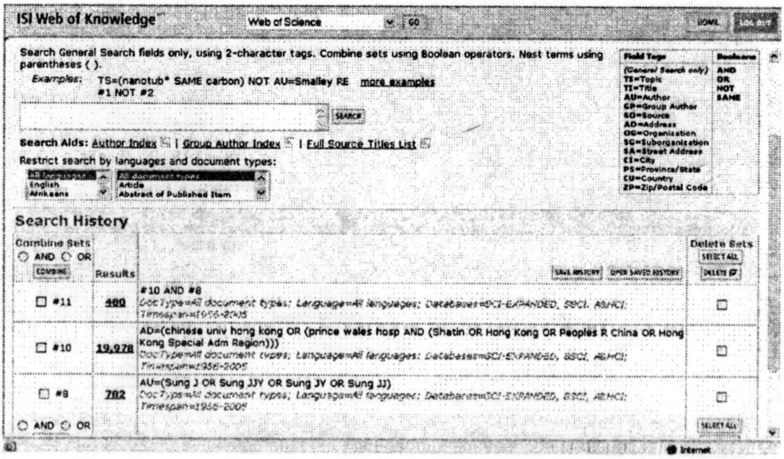

Conclusion

Now day's online resources provided various powerful search capability to end users. The accessibility of online resources are fast, reliable and round the clock. It increases the efficiency of library services and research productivity of the institutions/ universities. Banaras Hindu University library organized "Training-cum-User Education Programme" from time to time for effective use of its subscribed online resources. Library professionals are well aware of basic search techniques, but there is a need to update themselves with newly emerging technologies to provide effective and efficient information service to the users.

References

1. **Jain, Vivekanand** (2011) Search Engines: Utility and efficiency. New Delhi, Shree Publishers.
2. Online union catalogue of Indian Universities. Available at http:// indcat.inflibnet.ac.in/indcat/
3. JSTOR Database. Available at http://www.jstor.org/
4. Web of Science. Available at http://apps.webofknowledge.com
5. ProQuest dissertations and thesis database. Available at http:// search.proquest.com/pqdtft/index
6. Google Advanced search: http://www.google.co.in/ advanced_search
7. Notess, Greg R (2003) A decade on the net. *Online* 27(2). Available at: http://www.infotoday.com/online/Mar03/onthenet.shtml
8. Search Engine Showdown : www.searchengineshowdown.com
9. Search engines definition from wikipedia. http://en.wikipedia.org/ wiki/Search_engine
10. Silverstein, C, *et al.* (1999) Analysis of a very large web search engine query log. *SIGIR Forum,* V.33 (1) pp.6-12.

4

Open Access Electronic Information Sources for Education and Research

Mohammad Nazim &
Mustaq Ahmad

Abstract

This paper described open access publishing model and highlighted its importance for researchers, educational institutions and libraries. This paper listed various open access reference sources, books collections, full-text databases of journals, directories of institutional repositories, citation databases, and statistical databases. Features of each source are also discussed.

Introduction

With the development of information communication technologies, a number of alternative strategies to the traditional scholarly publishing system have been evolved. Among these, Open Access (OA) model is venerable which promise to be extremely advantageous to peers everywhere, especially to those who have acute shortage of resources for purchasing scholarly literature. The impetus of OA was boosted by the Open Society Institute (OSI) in a small meeting convened in Budapest on December 1-2, 2001. The purpose of the meeting was to accelerate progress in the international effort to make research literature in all academic fields freely available on the Internet (OAIS, 2002). The first major international statement on OA, which includes a definition, background information and a list of signatories, is the Budapest OA Initiative (Hirtle, 2001. The other two leading statements are the Bethesda Statement on OA Publishing and the Berlin Declaration on OA to Knowledge in the Sciences and Humanities (Harnad, 2002). The conception of OA in these three statements, which is often called the BBB (Budapest, Bethesda and Berlin) definition, launched, inspired, and continues to guide the OA movement.

What is OA?

By OA, we mean the free, immediate, availability on the public Internet of those works which scholars give to the world without expectation of payment–permitting any user to read, download, copy, distribute, print, search or link to the full text of these articles, crawl them for indexing, pass them as data to software or use them for any other lawful purpose (ARL, 2005). Essentially it means the provision of free, immediate (upon publication), permanent access to research results for anyone to use, download, copy and distribute. What makes it possible is the internet and the consent of the author or copyright-holder. There are two models of providing OA literature:

- "Green OA" is provided by authors publishing in any journal and then self-archiving their postprints in their institutional repository or on some other OA website. Green OA journal publishers endorse immediate OA self-archiving by their authors.
- "Gold OA" is provided by authors publishing in an OA journal that provides immediate OA to all of its articles on the publisher's website.

Why OA?

"OA truly expands shared knowledge across scientific fields—it is the best path for accelerating multi-disciplinary

breakthroughs in research" ?—Open Letter to the US Congress signed by Nobel Prize winners.

OA is cost effective way to disseminate and use information. The goal of the OA movement is to make scholarly literature freely available in digital form worldwide with minimal restrictions in their use. OA is beneficial for the researchers, educational institutions and particularly for libraries.

For Researchers

- Increases readers' ability to find useful and relevant literature
- Increases the visibility, readership and impact of author's works
- Creates new avenues for discovery in digital environment
- Enhances interdisciplinary research
- Accelerates the pace of research, discovery and innovation

For Educational Institutions

- Contributes to core mission of advancing knowledge
- Democratizes access across all institutions – regardless of size or budget
- Increases competitiveness of academic institutions
- Enriches the quality of their education
- Ensures access to all that students need to know, rather what they (or their school) can afford
- Contributes to a better-educated workforce

For Libraries

OA is an alternative to the traditional subscription–based publishing model possible by new digital technologies, networked communication and availability of information sources on public domain.

OA Electronic Information Sources

A study published in 2010 showed that of the total output of peer-reviewed articles roughly 20 % could be found Openly Accessible (Haider, 2010). 8.5 % of the journal literature could be found free at the publishers' sites ("Gold OA"), of which 62 % in full OA journals, 14 % in subscription journals making their electronic versions free after a delay, and 24 % as individually open articles (against payment) in otherwise subscription journals. For an additional 11,9 % of the articles free full text copies were found elsewhere ("Green OA") in either subject-based repositories (43 %), institutional repositories (24%) or on the home pages of the authors or their departments (33%). These copies were further classified into exact copies of the published article (38 %), manuscripts as accepted for publishing (46 %) or manuscripts as submitted (15 %).

There is variety of sources available freely over Internet, which may be grouped into the following categories:

1. Reference sources (dictionaries, thesaurus and encyclopedias)
2. Books collections
3. Full text journals
4. Institutional Archives/Repositories
5. Citation Databases
6. Statistical databases

Reference Sources

The Free Dictionary : This is an online dictionary (Houghton Mifflin), thesaurus (Collins) and encyclopaedia (Columbia), with English and American audio pronunciation, translation to other languages and other features such as word games.

Wikinitionary : This is the sister project of Wikipedia which provide meanings and uses of millions of words. Like Wikipedia, Wikinitionary also provides the facility to add words, meanings and their uses by anyone and that is why it called user generated dictionary.

Cambridge Dictionaries online : This site provides the facility to get all the words you need at your fingertips

Thesaurus.com : This site uses Roget's Interactive Thesaurus to provide synonyms for words and phrases.

Wikipedia : A free online encyclopedia that is written collaboratively by people from around the world. It contains over 2.8 million articles in over 100 languages.

Virtual Reference Collection : MIT's virtual reference collection contains links to web sites for items usually found in a library's general reference collection.

OneLook : This service will simultaneously search online dictionaries and glossaries worldwide for definitions. The collection includes both general and specialist reference sources, including works for art, business, computing, medicine, science, technology and slang.

***ROUTES* (http://routes.open.ac.uk/) :** ROUTES is a database providing access to selected quality-assessed freely available internet resources, selected by course teams and the Open University Library's Learning and Teaching Librarians.

Book Collections

***Electronic literature collection* (http://collection.elitera ture.org/) :** An extensive database of listings for electronic works, their authors, and their publishers. The descriptive entries cover poetry, fiction, drama, and non-fiction that make significant use of electronic techniques or enhancements.

***Project Gutenberg* (http://www.gutenberg.org/wiki/ Main_Page) :** The purpose of Project Gutenberg is to digitize important books in the field of literature, science and humanities. Now it has a collection of over 36,000 freely-available copyright expired e-books, in a wide selection of languages and searchable, with list of "top" authors and titles.

***Bibliomania* (http://www.bibliomania.com/) :** This is an American site giving access to thousands of free electronic books, poems, articles, short stories and plays. It also includes study guides, dictionaries, biographies, religious texts and popular non-fiction.

***UC Press e-books Collection 1982-2004* (http://publishing.cdlib.org)** : More than 300 free electronic editions published by eScholarship Editions, mainly in the humanities, religion, history, literature, arts and social sciences.

Full Text Journals

***Directory of OA Journals (DOAJ)* (http://www.doaj.org)** : DOAJ covers free, full text, quality controlled scientific and scholarly journals, aiming to cover all subjects and languages. As of March 2011, there are 2605 journals in the directory.

***African Journals On Line (AJOL)* (http://www.ajol.info)** : This site provides access to citations and full text of over 261 African journals covering almost all subject areas. AJOL also offers a document delivery service which is free to users of developing countries.

***Highwire* (http://highwire.stanford.edu/lists/freeart.dtl)** : Highwire facilitates access to a large archive containing 1432 free, science based full-text journals with over 2.5 million full text scholarly articles on medical/biomedical topics. Most journal titles include back issues older than 12-24 months.

Institutional Archives/Repositories

***Directory of OA Repositories (OpenDOAR)* (http://www.opendoar.org)** : This directory produced by University of Nottingham, OpenDOAR is a directory of academic OA repositories located throughout the world, which have been quality assessed. As well as providing a simple repository list. OpenDOAR provides the facility to search contents separately for individual repositories, as well as search contents across the whole range of repositories. At the end of March 2011 OpenDOAR listed over 1800 repositories

***OAIster* (http://oaister.umdl.umich.edu/cgi/b/bib/bib-idx?c=oaister;page=simple)** : OAIster is a collection of freely available, previously difficult-to-access, academically-oriented full-text resources searchable without restriction. OAIster includes over 9 million records from over 600 institutions worldwide.

ARC-A Cross Archive Search **Service (http://arc.cs.odu.edu) :** ARC is an experimental research service of Digital Library Research group at Old Dominion University. ARC searches across over 278 international repositories through a unified search interface.

ePrints **(http://eprints-uk.rdn.ac.uk/search/?view=advanced) :** ePrints aims to provide national, discipline-focused searching for access of journal articles, technical reports and web pages in electronic institutional archives of 41 selected UK universities and colleges.

Citation Databases

Scirus **(http://www.scirus.com/) :** Scirus is the most comprehensive scientific research tool on the web. With over 410 million scientific items indexed at last count, it allows researchers to search for not only journal content but also scientists' homepages, courseware, pre-print server material, patents and institutional repository and website information. It allows researchers to search for not only journal content but also scientists' homepages, courseware, pre-print server material, patents and institutional repository and website information.

Google Scholar **(http://scholar.google.co.in/) :** Google Scholar provides a simple way to broadly search for scholarly literature. From one place, you can search across many disciplines and sources: articles, theses, books, abstracts and court opinions, from academic publishers, professional societies, online repositories, universities and other web sites. Google Scholar helps you find relevant work across the world of scholarly research. Google Scholar aims to rank documents the way researchers do, weighing the full text of each document, where it was published, who it was written by, as well as how often and how recently it has been cited in other scholarly literature.

Open-JGate : Open-JGate is electronic database which index articles of OA journals. This database is created maintained by

Infomatics India Pvt. Ltd. This database provides the facility of search by author, keywords, subject, etc.

LISTA (Library, Information Science and Technology Abstract) : LISTA is also an OA databases covers more than 200 journals in the field of library science, information science, computer science, information technology. LISTA is provided by reputed information product supplier named EBSCO.

***JOLIS Library Catalog / World Bank/IMF* (http://jolis.worldbankimflib.org/e-nljolis.htm) :** The Jolis Library Catalog is the catalog of the IMF/World Bank Library Network. The catalog, which contains over 1.2 million items includes references to a wide variety of development related materials from hundreds of different publishers. The catalog also includes references, and links to many published IMF and World Bank materials.

***UNESBIB - UNESCO Documents Database* (http://unesdoc.unesco.org/ulis) :** UNESBIB includes over 100,000 citations for books, articles and UNESCO publications, some with full text links. Languages included are: English, French, Spanish, Arabic and Russian.

***PubMed* (http://www.ncbi.nlm.nih.gov/entrez) :** PubMed provides access to over 11 million citations from the U.S. National Library of Medicine and other related databases. Links to selected online journals, some freely available, are also included.

***Popline* (http://db.jhuccp.org/popinform/basic.html) :** Popline is the world's largest bibliographic database on population, family planning, and related health. Citations also cover sexually transmitted diseases including HIV/AIDS, reproductive health, law, and policy issues. The database includes abstracts of journal articles, monographs, technical reports, and unpublished works.

***UNBISnet / U.N. Dag Hammarskjold Library* (http://unbisnet.un.org) :** Catalogue of United Nations(UN) documents and publications indexed by the UN Dag Hammarskjöld Library and the Library of the UN Office at Geneva. Also included are commercial publications and other non-UN sources held in the collection of the Dag Hammarskjöld Library. The coverage of UNBISnet is from 1979 onward, however, older documents are being added to the catalogue on a regular basis as a result of retrospective conversion. UNBISnet also provides instant access to a growing number of full text resources in the six official languages of the UN (Arabic, Chinese, English, French, Russian and Spanish), including resolutions adopted by the General Assembly, the Economic and Social Council and the Security Council from 1946 onward.

Statistical Data Sources

World Development Indicators **(http://www.worldbank.org/data)** : World Development Indicators (WDI) is the World Bank's annual compilation of data about development. The WDI includes more than 800 indicators in 83 tables organized in 6 sections: World View, People, Environment, Economy, States and Markets, and Global Links. Data are shown for 152 economies with populations of more than 1 million and 14 country groups, plus selected indicators for 56 other smaller economies

FAOSTAT Database / Food and Agriculture Organization **(http://faostat.fao.org)** : Multilingual statistical databases containing over 1 million time-series records covering international statistics in the areas of production, trade, food balance sheets, fertilizer and pesticides, land use and irrigation, forest products, fishery products, population, agricultural machinery, and food aid shipments.

LABORSTA Database / International Labour Organization **(http://laborsta.ilo.org)** : Contains yearly statistics of employment, unemployment, hours of work, wages, labor cost, consumer price Indices, occupational injuries, strikes and lockouts on over 200 countries (data since 1969); monthly statistics of employment, unemployment, hours of work, wages, consumer price indices (data since 1976); and economically active population estimates and projections, 1950-2010.

UN Monthly Bulletin of Statistics / United Nations **(http://unstats.un.org/unsd/mbs)** : Includes current monthly economic statistics for most countries and areas of the world. The statistics are obtained by from official sources in the various countries, except where otherwise stated in the notes to the tables. Updated monthly.

UNSTATS UN Common Database / United Nations **(http://unstats.un.org/unsd/cdb) :** Draws selectively on statistics from throughout the UN system, covering all countries, areas and over 300 series from more than 30 specialized international data sources. Time series data is generally available from 1970 or 1980. Many series are disaggregated to show underlying distributions. The source includes comprehensive footnotes and meta-information on sources, definitions, and frequency of updates, and provides technical definitions and standards verbatim from their original sources. Users may view data, compile graphs, calculate derived measures, and export data.

Conclusion

Due to financial crises, reduction of annual budget of libraries and ever increasing prices of books and other information sources, libraries are no longer in a position to continue the subscription of their existing information sources. Therefore, OA is an alternate to subscription-based information sources particularly to the institutions/libraries which have limited financial resources. But most of the researchers are not aware of availability of OA information sources in their subject area. Libraries and librarians could play significant roles by organizing awareness programs and listing resources subject-wise on library website.

References

1. *Association of Research Libraries* (2005). New Paradigms: The OA Movement. Available http://www.arl.org/sc/models/oa.shtml (Last accessed on 12.4.2011).
2. **Haider, J** (2008). *The Geographic Distribution of OA Journals*". Digital Library of Information and Technology. Available http://dlist.sir.arizona.edu/939/ (Last accessed on 10.04.2011).

3. **Harnad, S.** (2002). *The Self-archiving Initiative. Nature: Web debates*. Available: http://www.nature.com/nature/debates/e-access/Articles/harnad.html.
4. **Hirtle, Peter** (2001). *OAI and OAIS: What's in a name? D-Lib Magazine*, 7(4), April, 2001. Available at: http://www.dlib.org/dlib/april01/04editorial.html.
5. **OAIS** (2002). *Reference model for an Open Archival Information System (OAIS)*, Consultative Committee for Space Data Systems, CCSDS 650.0-B-1, Blue Book, Issue 1, January, adopted as ISO 14721:2003. Available at: http://ssdoo.gsfc.nasa.gov/nost/wwwclassic/documents/pdf/ CCSDS-650.0-B-1.pdf.

5

Electronic Resources in Banaras Hindu University Library System : A Study

Anil Agrawal

Introduction

Electronic publishing has been revolutionizing the format of recorded knowledge. Electronic information services are attracting the users attention in todays networked environment. Electronic journals and databases has numerous advantages over its print counterpart. Apart from the fact that most of the print journals are expensive, there is rise in the subscription price of journals and databases on an exponential rate. Financial constraints because of static/shrinking grants available to higher education institutions have forced them to cut their subscriptions drastically and the libraries have to resort to alternatives like consortia. During the last few years, a number of consortium have come up in India and among them UGC-Infonet is one such Consortium available to Indian Universities.

Indian Universities constitutes one of the largest higher education system in the world comprising of approximately 400 universities and 17000 affiliated colleges. These institutions are facing acute shortage of funds to subscribe the journals and databases. With the formation of UGC-Infonet these institutions are able to provide access to number of electronic journals and databases to its users.

Scope of the Study

This study is confined to the availability of E-Resources like E-Journals, E-Databases and E-Books in the Banaras Hindu University Library System which ranks among the top five University in the priority list of Information and Library Network system (INFLIBNET) of the country.

Banaras Hindu University Library System

The Banaras Hindu University Library system, the largest University Library System in the country. Beginning with a small but precious donated collection, the library grew by leaps and bounds with magnificent donations of personal and family collections from many eminent personalities and families like Lala Sri Ram of Delhi, Jamnalal Bajaj of Wardha, Roormal Goenka, Batuk Nath Sharma, Tagore Family collection, Nehru Family collection, etc. amongst a score of others and purchase of books and journals out of the regular fund with the result that it has a collection of around 60,000 volumes in 1931 itself. The trend of donation of personal and family collection to the library continued as late as forties with the result that it has unique pieces of manuscripts, rare books and journals dating back to 18th century.

With this sound footings and background, the library took long strides during sixties and seventies in its development and metamorphosed in a system of libraries with the establishment of institute, faculty and departmental libraries during the period. Presently the Banaras Hindu University Library System consists of Central Library at apex and 4 Institute Libraries, 16 Faculty Libraries, 25 Departmental Libraries, with a total collection of over 13 lakh volumes to serve the students, faculty members, researchers, and non techning officers and staff of sixteen faculties consisting of 140 subject departments of the university.

4.0 Collection of The BHU Library System

BHU Library system has a very rich collection of Print and Electronic documents. The collection of the BHU Library system are as follows :

Books	9,40,541
Journals (Bound Vols)	1,37,859
Current Journals	649
Ph.D Theses	12,046
Manuscripts	7,233
UN & Govt. Publications, Staff Publications, Rare & Out of Print Books, Local History Collection, University & its Founder Collection	3,500
Online Journals	9,786
Databases	42
E-Books	20,278

Source : *BHU Library website (http://internet.bhu.ac.in/bhulibrary/services.htm)*

Available e-Resources

E-resources are available to the library either through UGC-Infonet or through the payment of subscription fee. The library provides access to these electronic resources (E-Journals, E-databases and E-Books) to the users through campus wide network.

E-Journals

American Chemical Society (ACS) : The American Chemical Society is a self-governed individual membership organization that consists of more than 163,000 members at all degree levels and in all fields of chemistry. The organization provides a broad range of opportunities for peer interaction and career development, regardless of professional or scientific interests. The library provides access to the 37 online full text journals covering the subject fields like : Biochemical Research Methods, Biochemistry, Biochemistry Biotechnology, Applied Microbiology, Chemistry-Analytical, Applied, Inorganic and Nuclear, Medicinal, Organic, Physical, Crystallography. URL: http://www.pubs.acs.org/

American Institute of Physics (AIP) : The American Institute of Physics (AIP) is a not-for-profit membership corporation chartered in New York State in 1931 for the purpose of promoting the advancement and diffusion of the knowledge of physics and its application to human welfare. It is the mission of the Institute to serve physics, astronomy, and related fields of science and technology by serving its Member Societies and their associates, individual scientists, educators, students, R&D leaders. The library provides access to 18 full text electronic journals. URL: http://www.aip.org

American Physical Society (APS) : American Physical Society publishes the world's most prestigious and widely-read physics research journals and aims to develop and implement effective programs in physics education. APS informs its members of the latest developments through APS News, Physical Review Focus, and articles in Physics Today and communicates with the public and policymakers via the national media and a public web site: www. physicscentral.com. The library provides access to 10 full text electronic journals. URL: http://publish.aps.org/browse.php

Annual Reviews : Annual Reviews offer comprehensive, timely collections of critical reviews written by leading scientists. Annual Reviews volumes are published each year for 30 focused disciplines within the Biomedical, Physical, and Social Sciences. Annual Reviews is a nonprofit organization whose mission is to provide the worldwide scientific community with a useful and intelligent synthesis of the primary research literature for a broad spectrum of scientific disciplines. Annual Reviews publications are among the most highly cited in scientific literature. All Annual Reviews series are ranked within the top ten publications for their respective disciplines. Each year, Annual Reviews critically reviews the most significant primary research literature. The library provides access to 33 full text online journals. URL : http://arjournals.annualreviews.org/

Cambridge University Press Journals : It is the oldest printing and publishing house in the world. Cambridge Journals Online (CJO) provides full text for over one hundred journals in the sciences, social sciences, and humanities. Important features includes simultaneous full-text access for all journals subscribed and a powerful search engine, which provides full text searching throughout the database, allows the user to set search parameters, and offers the advantages of boolean searching. The library provides access to 224 full text electronic journals. URL: http://journals.cambridge.org/

Institute of Physics (IOP) : The Institute of Physics is a leading international professional body and learned society, established to promote the advancement and dissemination of information in physics. It plays a major role in setting professional standards for physicists and awarding professional qualifications and promoting physics through scientific conferences, education and science policy advice. The subject range covers: Condensed Matter and Materials Science, Applied Physics, Applied Mathematics and Mathematical Physics, Measurement Science and Sensors, Plasma Physics, Optical, Atomic and Molecular Physics, High Energy and Nuclear Physics, Medical and Biological Physics, Physics Education, Computer Science. BHU library provides access to 46 full text electronic journals. URL: http://iopsciece.iop.org/journals

JSTOR : JSTOR is a not-for-profit organization with a dual mission to create and maintain a trusted archive of important scholarly journals, and to provide access to these journals as widely as possible. The library provides access to 1401 full text electronic journals covering the subjects like Business and Management, Arts, History, Medical Science, Sociology, Area Studies, Literature, Language and Linguistics, Medical Science, Economics etc. URL : http://www.jstor.org/

Nature : Nature Publishing Group (NPG) aims to provide the world's premier information resource for the basic biological and physical sciences. NPG aims to communicate the latest ground-breaking and original scientific discoveries across all disciplines of science. The NPG journals subscribed by BHU library includes : Nature, Nature Biotechnology, Nature Genetics, Nature Immunology, Nature Materials, Nature Photonics, Nature Physics and Nature Review Microbiology. The subject area covered by these journals includes General Science, Chemistry, Life Sciences Physical Sciences and Physics. URL: http://www.nature.com/

Project Muse : Project MUSE is a unique collaboration between libraries and publishers, providing 100% full-text, affordable and user-friendly online access to a comprehensive selection of prestigious humanities and social sciences journals. Currently, Project MUSE(r) offers quality journal titles from 40 scholarly publishers as one of the academic community's primary electronic journals resources. Project MUSE covers the fields of literature and criticism, history, the visual and performing arts, cultural studies, education, political science, gender studies, economics, and many others. The library provides access to 411 electronic journals. URL: http://muse.jhu.edu/journals

Royal Society of Chemistry : The Royal Society of Chemistry (RSC) is the Professional Body for chemists and the Learned Society for chemistry. It is one of the most prominent and influential independent scientific organisation in Britain. Through its 46,000 members, including academics, teachers and industrialists, the RSC promotes the interests of chemists and the benefits of chemical science. BHU library provides access to 29 full text online journals and 6 databases. URL: http://www.rsc.org/Publishing/journals/

Elsevier's Science Direct : Science Direct is the web-based interface to the full-text database of Elsevier Science journals, one of the world's largest providers of scientific, technical and medical (STM) literature. The Science Direct offers a rich electronic environment for research journals, bibliographic databases and reference works. The library provides access to more than 1036 full text journals from 10 subject collections with backfiles from

1995 onwards covering the subjects like Biochemistry, Genetics & Mol. Biology, Agriculture & Biological Science, Chemistry, Computer Science, Economics, Immunology & Microbiology, Mathematics, Physics & Astronomy, Social Sciences, and Psychology along with an expanding suite of bibliographic databases and linking to another one million full-text articles via Cross Reference to other publishers' platforms. URL : http://www.sciencedirect.com/

Emerald Full-text : Emerald established in 1967 by a group of senior academics formed MCB University Press, a publishing house that focused on niche management disciplines including strategy, change management, and international marketing. Emerald publishes the world's widest range of management and library and information services journals, as well as a strong specialist range of engineering, applied science and technology journals. The library electronic databases allow instant access to the latest research and global thinking. It provides the information, ideas and the opportunity to gain insight into the key management topics. The library provides access to over 200 electronic journals covering the subjects of Management and Library and Information Science. URL : http://www.emeraldinsight.com

Oxford University Press : Oxford Journals is a division of Oxford University Press, which is a department of Oxford University. It publishes well over 230 academic and research journals covering a broad range of subject areas, two-thirds of which are published in collaboration with learned societies and other international organizations. The library provides access to 206 online journals from the disciplines of Economics, Humanities , Law, Life Sciences, Mathematics & Physical Sciences, Medicine and Social Sciences. URL : http://www.oxfordjournals.org

Sage E-Journals : Founded in 1965, SAGE is the world's leading independent academic and professional publisher .The library provides access to around 200 journals from Sage covering the subjects like Education, Management & Organization Studies, Psychology, Sociology, Political Science. URL : http://online.sagepub.com/

Wiley Blackwell : Wiley-Blackwell is the international Scientific, Technical, Medical and Scholarly publishing business of John Wiley & Sons, with strengths in every major academic and professional field and partnerships with many of the world's leading societies. Wiley-Blackwell publishes 1500 peer-reviewed journals and 1500+ new books annually in print and online, as well as databases, major reference works and laboratory protocols. The library provides access to 908 titles covering the subjects like Accounting and Finance, Archeology, Chemistry, Crystallography, Materials Science, Economics, Agricultural Science, Anthropology, Law etc. URL : http://onlinelibrary.wiley.com/

Springer Link : Springer develop, manage and disseminate knowledge – through books, journals and the Internet. The library provides access to 1389 Journals covering the fields like Architecture and Design, Behavioral Science, Biomedical and Life Sciences, Business and Economics, Chemistry and Materials Science, Computer Science, Earth and Environmental Science, Engineering, Humanities, Social Sciences and Law, Mathematics and Statistics, Medicine, Physics and Astronomy. URL : http://www.springerlink.com/

Taylor & Francis : Taylor & Francis Group is an Informa business (www.informa.com). Taylor & Francis Group publishes more than 1,500 journals and around 1,800 new books each year. The library provides access to 1365 journals covering the fields like Agriculture & Environmental Sciences, Arts, Behavioral Sciences, Computer Science, Earth Sciences, Economics, Education, Engineering & Technology, Humanities, Law. URL : http://www.tandf.co.uk/journals/

5.2 E-Databases

MathSciNet : MathSciNet is a comprehensive database covering the world's mathematical literature since 1940. It provides Web access to the bibliographic data and reviews of mathematical research literature contained in the Mathematical Reviews Database. The MathSciNet has signed reviews, powerful search functionality, and timely updates. It fosters the navigation of mathematics literature by providing links to original articles and other original documents, when available, and by encouraging links from journal article references to MathSciNet. The MathSciNet offers free access to Featured Reviews, those reviews from the Mathematical Reviews database that were especially commissioned for some of the books and papers that are considered particularly important in the areas that they cover.

Web of Science : The ISI Web of Science provides access to information for all levels of academic, corporate, and government research. It offers a comprehensive, fully integrated platform that empowers researchers and accelerates discovery. It offers citations and cited reference searching. The ISI Web of Knowledge provides a single interface, enabling natural-language searches across multiple content sources: journal articles; proceedings papers; patents; chemical reactions and compounds; and content from preprint, funding information, and research activity websites.

SciFinder Scholar : SciFinder Scholar is a Z39.50 Windows-based interface that provides easy access to the rich and diverse scientific information contained in the CAS databases including Chemical Abstracts from 1907 onwards. The SciFinder Scholar offers a variety of pathways to explore CAS databases as well as MEDLINE. SciFinder Scholar interface provides the most accurate and comprehensive chemical and related scientific information including: journal articles and patents together in one source, substance data, chemical reactions, chemical regulatory data, chemical suppliers, biomedical literature. SciFinder Scholar covers subjects like Chemistry and also Agriculture, Biology and Life Sciences, Engineering, Food, Geology, Medicine, Physics, Polymer and Material Sciences.

ISID : The Institute for Studies in Industrial Development (ISID), is a national-level policy research organization in the public domain and is affiliated to the Indian Council of Social Science Research (ICSSR). ISID has developed databases on various aspects of the Indian economy, particularly concerning industry and the corporate sector. It has created On-line Indexes of Indian

Social Science Journals (OLI) and Press Clippings on diverse social science subjects. These have been widely acclaimed as valuable sources of information for researchers studying India's socio-economic development.

ProQuest Dissertations and Theses : This database is the world's most comprehensive collection of dissertations and theses from around the world, spanning from 1861 to the present day and offering full text for most of the dissertations added since 1997. The subject area covered includes : Business, Health & Medicine, History, Literature and Language, Science and Technology, Social Sciences and Arts.

5.3 E-Books

Cambridge University Press Collection : The library provides access to more than 500 e-books on the subjects like: Micro and Macro Economics, International Economics, Political Economics & Finance, Physics, Language & Linguistics, Business & Management, Cambridge Companions Online, Cambridge Histories Online, Shakespeare Survey Online etc from Cambridge University press.

Encyclopedia Britannica : The library provides access to full text articles from the Encylcopedia Britannica.

Sage- E -Books Collections : Sage Publication is one of the renowned publisher in the world. The library provides online access access to 47 titles of Encyclopedia and Handbooks viz. Encyclopedia of Activism and Social Justice, Encyclopedia of African American Society, Encyclopedia of Anthropology and Encyclopedia of Behavior Modification and Cognitive Behavior Therapy etc. from Sage publication.

Springer Book Collections : The library provides access to more than 11000 e-books from Springer Books covering the subjects like Architecture and Design, Behavioral Science, Biomedical and Life Sciences, Business and Economics, Chemistry and Materials, Science, Computer Science, Earth and Environmental Science, Engineering, Humanities, Social Sciences and Law, Mathematics and Statistics.

Conclusion

The university library system in India faces financial problems to satisfy the users need. But, the efforts of Inflibnet through UGC-Infonet Consortium are appreciable and will definitely strengthen higher education system in India. The free access to scholarly online resources will help educational institutions in translating their mission into reality. The research output will increase multifold. The Banaras Hindu University Library system is trying its level best by providing access to the online journals to its users through UGC-Infonet as well as subscribing e-journals, e-books, database according to the users demand. But, the library still find it very difficult to satisfy all the users community.

References

1. **Chakravarty, Rupak and Singh, Sukhwinder** (2005). E-Resources for Indian universities : New initiatives. *SRELS Journal of Information Management* Vol. 42, No. 1, March 2005, Paper D. p57-73.
2. **Kahiser, Nikam and B. Pramodini** (2007). Use of e-journals and databases by academic community of University of Mysore : A survey. *Annals of Library and Information Studies*. Vol. 54, pp. 19-22
3. http://arjournals.annualreviews.org/
4. http://internet.bhu.ac.in/bhulibrary/history.htm
5. http://internet.bhu.ac.in/bhulibrary/services.htm
6. http://iopsciece.iop.org/journals
7. http://journals.cambridge.org/
8. http://muse.jhu.edu/journals
9. http://onlinelibrary.wiley.com/
10. http://online.sagepub.com/
11. http://publish.aps.org/browse.php
12. http://www.aip.org
13. http://www.bhulibrary.ac.in

14. http://www.emeraldinsight.com
15. http://www.jstor.org/
16. http://www.nature.com/
17. http://www.oxfordjournals.org
18. http://www.pubs.acs.org/
19. http://www.rsc.org/Publishing/journals/
20. http://www.sciencedirect.com/
21. http://www.springerlink.com/
22. http://www.tandf.co.uk/journals/

6

Free Access to Law Resources via Legal Information Institutes

Ram Kumar Dangi &
R.P. Bajpai

Abstract

It deals with legal information resources and their availability to all with the Montreal Declaration on public access to law or free access to law movement. It also emphasis on effective legal initiatives in USA, UK and India with some legal institutes.

Keywords: Legal Information, Free access to law, WorldLii, BaiLii, Lii (cornell), IndLii.

Introduction

Public legal information means legal information produced by public bodies that have a duty to produce law and make it public. It includes primary sources of law, such as legislation, case law and treaties, as well as various secondary (interpretative) public sources, such as reports on preparatory work and law reform, and resulting from boards of inquiry. It also includes legal documents created as a result of public funding. Publicly funded secondary (interpretative) legal materials should be accessible for free but permission to republish is not always appropriate or possible. In particular free access to legal scholarship may be provided by legal scholarship repositories, legal information institutes or other means.

Free Access to Law Movement

The Free Access to Law Movement is the umbrella name for the collective of legal projects across several common law countries to provide free online access to legal information such as case law and legislation. The movement began in 1992 with the creation of the Cornell Law School Legal Information Institute by Tom Bruce and Peter Martin. The name Legal Information Institute has been widely adopted by other projects. It is usually prefixed by a country or region identifier

Declaration on Free Access to Law

In October 2002 the meeting of LIIs in Montreal at the 4th Law via Internet Conference, made the following declaration as a joint statement of their philosophy of access to law. There were some further modifications of the Declaration at the Sydney meeting of LIIs in 2003 and at the Paris meeting in 2004. Legal information institutes of the world, meeting in Montreal, declare that:

- Public legal information from all countries and international institutions is part of the common heritage of humanity. Maximizing access to this information promotes justice and the rule of law;
- Public legal information is digital common property and should be accessible to all on a non-profit basis and free of charge;
- Independent non-profit organizations have the right to publish public legal information and the government bodies that create or control that information should provide access to it so that it can be published.

A legal information institute publishes via the internet public legal information originating from more than one public body; provides free, full and anonymous public access to that information;

does not impede others from publishing public legal information; and supports the objectives set out in this Declaration.

All legal information institutes are encouraged to participate in regional or global free access to law networks and the legal information institutes agree on the following:

- To promote and support free access to public legal information throughout the world, principally via the Internet;
- To cooperate in order to achieve these goals and, in particular, to assist organisations in developing countries to achieve these goals, recognising the reciprocal advantages that all obtain from access to each other's law;
- To help each other and to support, within their means, other organisations that share these goals with respect to,
  - Promotion, to governments and other organisations, of public policy conducive to the accessibility of public legal information;
  - Technical assistance, advice and training;
  - Development of open technical standards;
  - Academic exchange of research results.
- To meet at least annually, and to invite other organizations who are legal information institutes to subscribe to this declaration and join those meetings, according to procedures to be established by the parties to this Declaration.
- To provide to the end users of public legal information clear information concerning any conditions of re-use of that information, where this is feasible.

World Legal Information Institute (WORLDLII)

The World Legal Information Institute is the umbrella project for all the other LII projects. From the WorldLII website all other databases can be searched. The WorldLII is a free, independent and non-profit global legal research facility developed collaboratively by the following Legal Information Institutes and other organizations.

- Australasian Legal Information Institute (AustLII)
- British and Irish Legal Information Institute (BAILII)
- Canadian Legal Information Institute (CanLII)
- Hong Kong Legal Information Institute (HKLII)
- Legal Information Institute (Cornell) (LII (Cornell))
- Pacific Islands Legal Information Institute (PacLII)
- Wits University School of Law (Wits Law School)

The LIIs, meeting in Montreal in October 2002, adopted the Montreal Declaration on public access to law. WorldLII comprises three main facilities: Databases, Catalog and Web search.

WorldLII Databases

WorldLII provides a single search facility for databases located on the following Legal Information Institutes: AustLII;

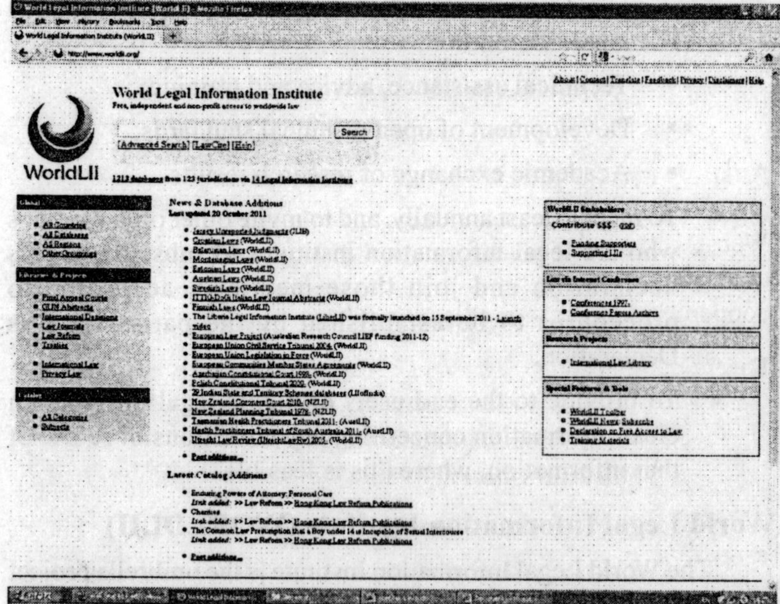

Fig. 1. An Electronic Site of World Legal Information Institute (http://www.worldlii.org/)

BAILII; CanLII; HKLII; LII (Cornell); and PacLII. WorldLII also includes as part of this searchable collection its own databases not found on other LIIs. These include databases of decisions of international Courts and Tribunals, databases from a number of Asian countries, and databases from South Africa (provided by Wits Law School). Over 270 databases from 48 jurisdictions in 20 countries are included in the initial release of WorldLII. Databases of case-law, legislation, treaties, law reform reports, law journals, and other materials are included. WorldLII welcomes enquiries concerning the possible inclusion of other databases on WorldLII or on one of its collaborating LIIs.

WorldLII Catalog and Websearch

The WorldLII Catalog provides links to over 15,000 law-related web sites in every country in the world. WorldLII's Websearch makes searchable the full text of as many of these sites as WorldLII's web-spider can reach. WorldLII welcomes enquiries from law librarians and other legal experts who are interested to become Contributing Editors to the WorldLII Catalog.

Operation of WorldLII

The provision of the WorldLII service is coordinated by the Australasian Legal Information Institute (AustLII), which maintains WorldLII's user interface, the WorldLII Catalog and Websearch, and the databases located only on WorldLII. Technical enhancements to WorldLII are being developed jointly by the cooperating Legal Information Institutes.

British and Irish Legal Information Institute (BAILII)

The British and Irish Legal Information Institute (pronounced 'Bailey') is the project providing legal information on England and Wales, Ireland, Northern Ireland, Scotland, United Kingdom, and the European Union. It was set up after a long and hard campaign by barrister Laurie West-Knights QC, Lord Saville and Lord Justice Brooke, who were concerned about the lack of availability of court judgments to ordinary court users and were inspired by the Australian LII.

The importance and volume of British case law in particular, coupled with initial political scepticism, made the achievement of BAILII all the more remarkable. It is now a major international resource of enormous value.

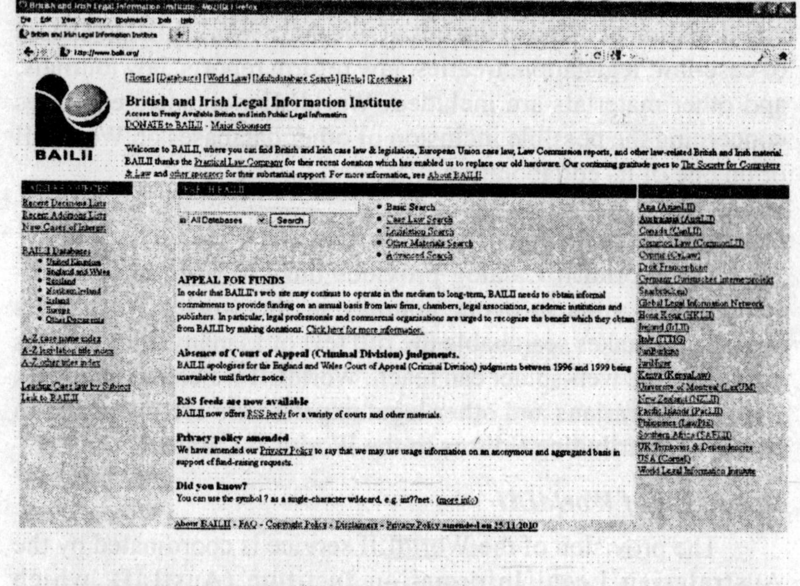

Fig. 2. An Electronic Site of British and Irish Legal Information Institute (www.bailii.org/)

Legal Information Institute at the Cornell Law School (LII)

It is the first Legal Information Institute, which was started at Cornell University Law School in 1992, and developed by 1994 a number of databases primarily of US federal law (particularly the US Code and US Supreme Court decisions). 'The LII', as it became known, was the first significant source of free access to law on the Internet, and demonstrated that a free access service could provide both high quality document presentation, and very high rates of access. The Institute provides free legal information for the United States. It was the original LII project. It is not-for-profit group that believes everyone should be able to read and understand the laws that govern them, without cost.

The Legal Information Institute at the Cornell Law School provides following legal resources i.e. Federal law, U. S. Constitution, U.S. Code, C.F.R., Supreme Court, Federal Rules, State law resources, State statutes , U.C.C., Uniform laws, World law, Wex legal encyclopedia, CRS Annotated Constitution, LII Supreme Court Bulletin etc.

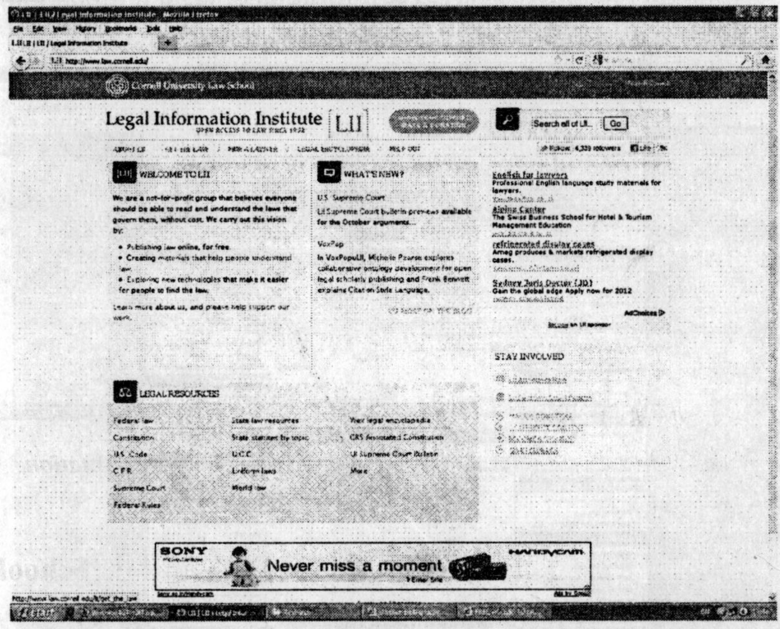

Fig. 3. ***An Electronic Site of Legal Information Institute at the Cornell Law School (http://www.law.cornell.edu/)***

India Legal Information Institute (INDLII), New Delhi, India

The First Free Legal Web Portal of India was inaugurated by the Prime Minister of India Dr. Manmohan Singh on the eve of

Law Day in New Delhi on 25 November, 2006 at a glittering function held at Vigyan Bhavan. The portal is virtual extension of India Legal Information Institute and is available at www.indlii.org. It is committed to collect legal information about India for all available sources; publish the same on the Internet with free and full public access; grant rights to the public to use the legal resources without any restrictions; create awareness about the availability of free legal resources; remove hurdles coming in the way of providing free legal information; coordinate with others Institutions to explore sources & utilization of legal information.

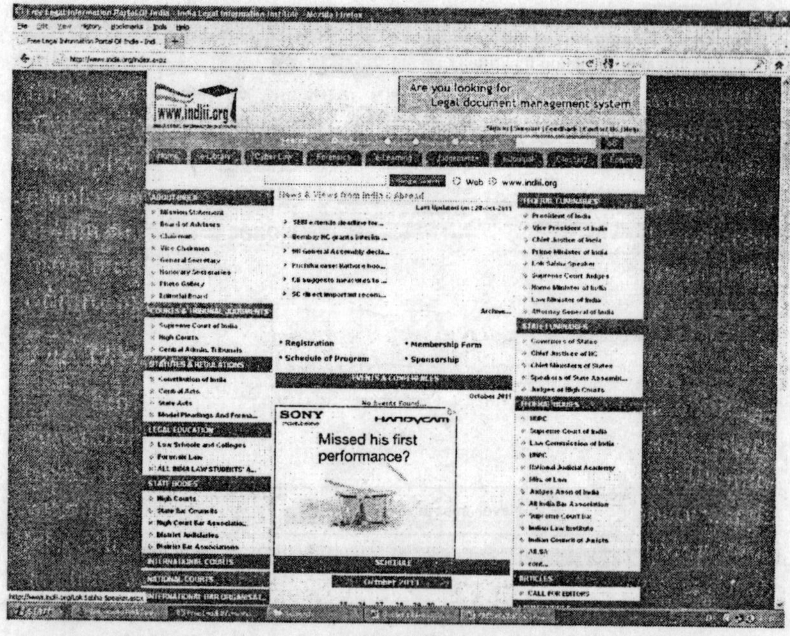

Fig. 4. An Electronic Site of India Legal Information Institute (http://www.indlii.org/index.aspx)

The portal provides information about central and state laws, judgments of various courts in the country besides news of the legal world. The Indian Judges and leaders of the Bar are actively associated with indlii. The idea behind the free portal is the mission of the Institute to provide free legal information to all. In the words of General Secretary of the Institute "everyone has a right to know law of the land free of cost and with this we started this initiative". The institute has a vast resource of highly qualified law experts who are dedicated to the cause of free flow of legal information.

The portal would play a great role in achieving this goal since anyone can get information from anywhere anytime. The portal is a complex platform of information about different aspects of law and judiciary in India with lot of references and cross references.

Indlii is an India based not for profit institute with a vision to make online and full public access to all publicly available legal information of India. The institute is committed to collect legal information about India for all available sources and create awareness about the availability of free legal resources.

Conclusion

Information for all is the main them behind all library movements worldwide. This is also main motto of IFLA world library congress. In the field of legal information, this movement initiated by Tom Bruce and Peter Martin in 1992 with the creation of the Cornell Law School Legal Information Institute to ensure that the law remains free and open to everyone and now it reaches to most of the countries of the world including India. These all efforts make general people to well inform citizen with their rights and duties.

References

1. **Carroll, Michael W.** (2006). *The Movement for Open Access Law.* Villanova University School of Law Working Paper Series Paper 50 retrieved from : http://law.bepress.com/villanovalwps/papers/art50.
2. **Greenleaf, Graham.** (2009). *The Global Development of Free Access to Legal Information Retrieved on 31.03.2011* from: http://austlii.edu.au/~graham/publications/2009/Greenleaf_LEFIS_Ch.pdf.

3. *India Legal Information Institute*. Available at http://www.indlii.org/index.aspx (visited 15 March 2011)
4. http://www.law.cornell.edu (visited 15 March 2011)
5. *World Legal Information Institute*. Available on http://www.worldlii.org/ (visited 15 March 2011)
6. *British And Irish Legal Information Institute*. Available on www.bailii.org/ (visited 15 March 2011)

7

Users Awarencess About E-Resources in Agriculture Sector : A Case Study of ICAR Libraries at Delhi

Harish Tripathi & Hans Raj

Abstract

Awareness is the primary key to utmost use of resources available in the library. Without User awareness it is very difficult to know about E-resources and knowledge retrieval tools are being used in the library. The present paper describes awareness of user's especially scientific community including scientists, technical officers, and research scholars in agricultural libraries, through conducting a survey of selected ICAR libraries at Delhi . The study reveals that majority of users (52%) prefer to use electronic resources for retrieving their required information, however (48%) user shown their interest in print resources. In this paper it is suggested that users awareness programmes must be started among scientific communities, for utmost use of electronic services in the library.

Keywords: *electronic resources; agricultural libraries; Delhi; ICAR; IARI; IASRI; NBPGR; NCAP online resources*

Introduction

There is no doubt, Internet or World Wide Web has changed the way of communication, collaborating and educating to users. Electronic resources and online resources, are being aligned in traditional collection in the libraries, in various format like CD-ROMs/Videos, e-Books, e-journals, online databases, consortium and internet or web resources. Most of publishers are bringing out their publications like encyclopedias, indexing and abstracting services, dictionaries, directories, year books, back volumes etc, in CD format/ online resources, with user friendly access and least cost, even sometimes free also. Consortia are the best method to buy and retrieve the e- information, with the effort of group of libraries. But it has come into notice that, In spite of having much e-resources /online resources, the nonuse or less use of them, is not good indication for library community as well as users community. So it is very necessary to make utmost awareness for optimum use of electronic information resources, by using various channel of communication formal and informal. Adequate collection of books, journals and other printing or electronic materials in ICAR libraries, fulfils the users needs to the great extent, by collaborating among the libraries, but the common quest of all libraries, the utmost use of available resources is not possible without awareness among users. The present study highlights the awareness about electronic resources among users, in ICAR libraries at Delhi, which are as under:

1. Indian Council of Agricultural Research (HQ) Library
2. Indian Agricultural Research Institute Library
3. Indian Agricultural Statistical Research Institute Library
4. National Bureau of Plant Genetic Research Library
5. National Centre for Agricultural Economics and Policy Research Library

Objectives

The main objectives of the present study are to make an assessment of the user awareness about e-resources in the ICAR libraries at Delhi. Thus the study is concerned with the following objectives.

- To assess the availability of e-resources in ICAR Libraries at Delhi.
- To find out the level of users awareness about library facilities including CeRA, NAIP project of ICAR.
- To find out e-resources which have a much or less demand
- To find out materials of e-resources which are used more frequently

Survey Methodology

The methodology used for collecting data is questionnaire as well as interview, as questionnaire was prepared and information were collected by personal visit and filled on the basis of discussion with users. User Community (no. 25 users) including Scientist, technical officers, Research Scholars, students, were asked the questions and data were collected, to find out the following objectives:

1. Whether you preferred to printed document or electronic resources/ free websites over Internet, for seeking your publication/ information.
2. How much awareness, about the following electronic resources available in the library.
   - Online Bibliographical database, like AGRICOLA, AGRIS etc
   - CeRA (Consortium for electronic Resources in Agriculture)
   - Krishi Prabha (Full text database of dissertations and PhD thesis submitted by students in Agricultural Universities)
   - Free agricultural/ general websites
   - CD ROMs/ DVDs
   - Resource sharing Network DELNET
3. How frequently used these e-resources by the user

E-resources about Electronic resources at ICAR Libraries at Delhi

E-resource refers to the information that requires computer access whether it is local accessed or accessed remotely through Internet. Therefore e-resources includes CD-ROMs, online journals, free/commercial online database, national and international websites, Resource sharing Networks, Consortia, Internet and so on. Information is provided to users in print format as well as electronic formats through OPAC facility of Library management Software and other electronic resources which are generally used in the libraries.

Table 1 reveals that CeRA accessible facility is available in all ICAR Libraries at Delhi. Apart from CeRA, IARI, IASRI, NBPGR and NCAP Library have 56, 40. 35 and 3 online journals respectively, however no online journal, is being subscribed in ICAR Library. CD Rom availability at Delhi Libraries of IASRI, NBPGR and NCAP Libraries are 629, 15 and 64 respectively. Except ICAR HQ Library, all ICAR Libraries at Delhi have the facility of full text accessibility of Krishi Prabha online. IARI library is very rich in online databases and online journals. Five online databases including Web of science and SCIRUS, and 56 online full text journals are being subscribed in IARI library. At present, IARI Library has completely moved from CD-ROM subscription to Online subscription.

Table no. 2 shows that in ICAR libraries at Delhi, 52% users are having their interest in electronic resources where as 48% readers still are having the interest in print resources. IARI library users and IASRI library users have their interest 84% and 60% respectively, in electronic resources, where as users of ICAR, users of NCAP and users of NBPGR Library, shown their interest 64%, 60% and 60% respectively, in print resources.

Users prefer to use both print and electronic resources available in library, But nowadays in both cases, electronic resources are brought in use to find out their required information, since Library management software and Resource sharing Network,

Table 1 : Availability of e-Resources in ICAR Libraries in Delhi

E-resources Library	*CD-ROMs*	*Online Journals*	*Online Database*	*CeRA*	*Krishi Prabha*	*LMS-Web-OPAC*	*Resource Sharing Network*
ICAR Library	35	00	No	Yes	No	E-Granthalaya	DELNET
IARI Library	#	56	5	Yes	Yes	Libsys	DELNET
IASRI Library	629	40	2	Yes	Yes	Alice for windows	DELNET
NBPGR Library	15	35	0	Yes	Yes	Libsys	No network
NCAP	64	3	1	Yes	No	Total Library Solution	No Network

Now IARI Library is subscribing online version of CD-ROMs and users access them through IP address.

Table 2 : user's Preference/interest About print v/s e-resources in ICAR Libraries at Delhi

Institutes	*Print resources*	*e-resources*	*Majority of Readers*
ICAR Library	16/25 (64%)	9/25 (36%)	Scientific and non-scientific community
IARI Library	04/25 (16%)	21/25 (84%)	Scientific community including Students,
IASRI Library	10/25 (40%)	15/25 (60%)	Scientific community, administrative officials
NBPGR Library	15/25 (60%)	10/25 (40%)	Scientific community & administrative officials
NCAP	15/25 (60%)	10/25 (40%)	Scientists, administrative officials
Total Average	48%	52%	

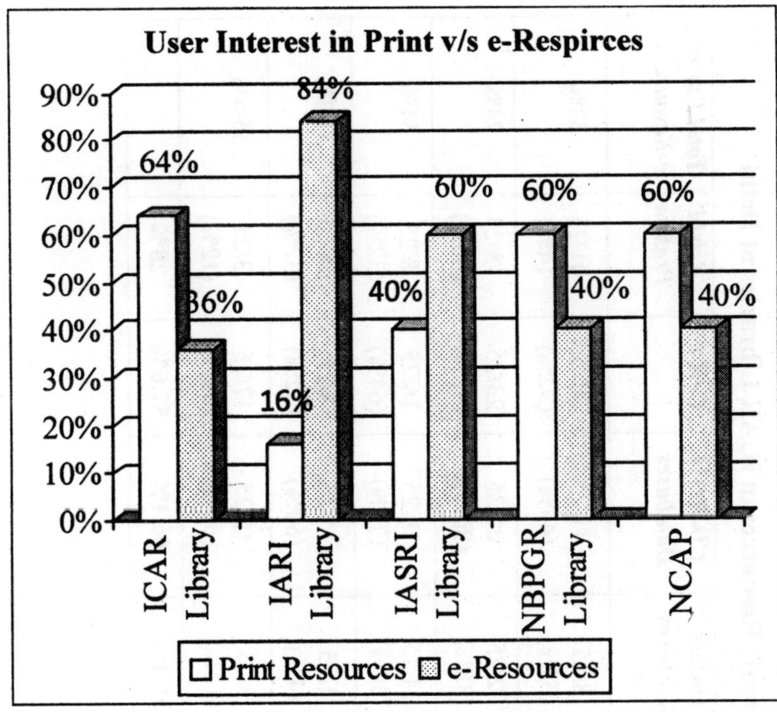

are also being brought in use to find out the required print publication available in library and outside the library. The study of users interest in printing materials and electronic material, have been showed in table no.2.

Table no. 3 indicates that 74% users of IARI , are well aware about of the use of e-resources, whereas level of awareness of IASRI users are 54% .. It is found that very less number of users 32% in ICAR HQ, 31.3% in NBPGR and 36.6% in NCAP are aware about these e-resources. Similarly awareness about the use of CD-Rom databases, LMS, and CeRA in ICAR Libraries at Delhi, was found 66.40%, 68.8% and 57.80 respectively, that can be said satisfactory, but awareness about Resource sharing Network, Online database and Krishi Prabha is 14.40%, 38.40% and 28% respectively, that is not much satisfactory report.

Table 3 : User Awareness of Electronic Resources in ICAR Libraries at Delhi

Institutes	*Use of OPAC/LMS by Users*	*Resource Sharing Networks*	*Online Databases*	*CD-ROM Databases*	*CeRA*	*Krishi Prabha*	*Total Use of e-Resources*
ICAR Library	15/25 (60%)	5/25 (20%)	8/25 (32%)	11/25 (44%)	8/25 (32%)	01/25 (4%)	32%
IARI Library	20/25 (80%)	7/25 (28%)	22/25 (88%)	24/25 (96%)#	23/25 (92%)	15/25 (60%)	74%
IASRI Library	19/25 (76%)	6/25 (24%)	14/25 (56%)	18/25 (72%)	16/25 (64%)	8/25 (32%)	54%
NBPGR Library	15/25 (60%)	0/25 (0%)	1/25 (4%)	15/25 (60%)	8/25 (32%)	08/25 (32%)	31.3%
NCAP Library	17/25 (68%)	0/25 (0%)	3/25 (12%)	15/25 (60%)	17/25 (68%)	3/25 (12%)	36.6%
Total average	**68.8%**	**14.4%**	**38.4%**	**66.4%**	**57.6%**	**28%**	

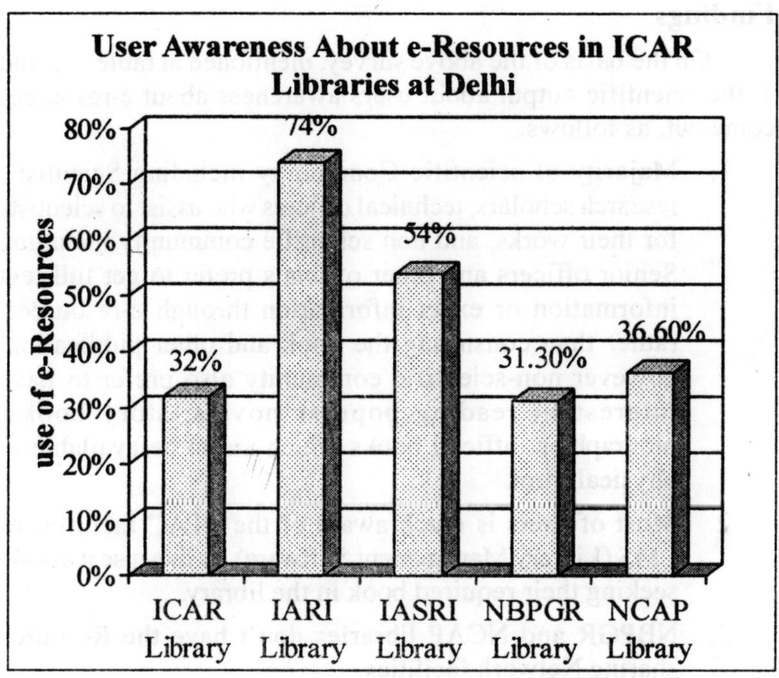
User Awareness About e-Resources in ICAR Libraries at Delhi
use of e-Resources
80%
70%
60%
50%
40%
30%
20%
10%
0%
32%
74%
54%
31.30%
36.60%
ICAR Library
IARI Library
IASRI Library
NBPGR Library
NCAP Library

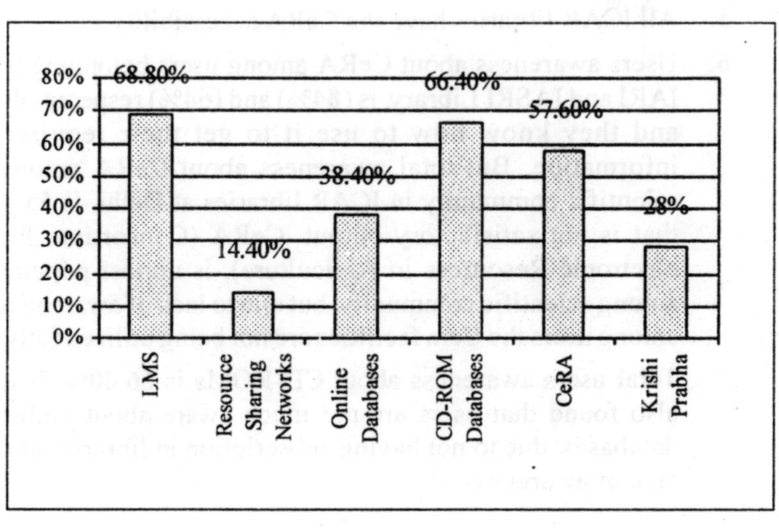
80%
70%
60%
50%
40%
30%
20%
10%
0%
68.80%
14.40%
38.40%
66.40%
57.60%
28%
LMS
Resource Sharing Networks
Online Databases
CD-ROM Databases
CeRA
Krishi Prabha

Findings

On the basis of the above survey, mentioned at table 1, 2 and 3, the scientific output about users awareness about e-resources come out, as follows.

1. Majority of scientific Community including Scientists, research scholars, technical officers who assist to scientists for their works, and non scientific community including Senior officers and other officials prefer to get full text information or exact information through e-resources, rather than consulting the book and other publication, however non-scientific community also prefer to have interest in reading popular novels, story books, biographies, official books which are to be available in physical form.
2. Most of users is much aware of the OPAC facilities of LMS (Library Management Software), as they use it during seeking their required book in the library.
3. NBPGR and NCAP libraries don't have the Resource sharing Network facilities.
4. Users awareness about Resource sharing Network (eg DELNET etc), is not satisfactory.
5. All ICAR libraries have the CeRA accessibility.
6. Users awareness about CeRA among users belonging to IARI and IASRI Library, is (84%) and (64%) respectively and they know how to use it to get their required information. But total awareness about CeRA among scientific community in ICAR libraries at Delhi, is 55% that is not satisfactory output. CeRA (Consortium for electronic Resources in Agriculture), is utmost popular among scientific community, but due to lack of awareness among users the Cera facilities are not being utilized fully.
7. Total users awareness about CD-ROMs is 66.40%. It is also found that users are not much aware about online databases, due to not having subscription in libraries and lack of awareness.

8. Besides IARI, Users awareness about 'Krishi Prabha' database is very less, in comparison with other e-resources.
9. It was found that IARI Library is very rich in print and e-resources but it can only be accessed via campus LAN/ Intranet, as it doesn't have its own site for general users, hence other users/libraries are not able to access resources available in library.
10. ICAR Library and NCAP Library don't have full text accessibility of Krishi Prabha

On the basis of the above-said facts and observation indicated in table no.1, 2 and 3, the final output of users' awareness about electronic resources available in the libraries, can be said good, but need to improve, in order to utmost of use of e-resources available in these libraries.

Suggestions

1. To improving user's awareness about information retrieval and facilities available in the library, awareness programmes should be organized by the each library frequently.
2. RSS feeds and e-alerts can be start as effective Current Awareness Service Tool in all these libraries. RSS (Rich Site Summary) feed and E-alerts help users about recent information. It can be used as additional library value added services via library website and delivered web content to users.
3. The suitable strategy should be adopted to promote e-resource, in order to increase the value of library in the organization and educate the scientific community about utmost use of e-resources.
4. The focus should be given to give orientation to scientific community on following points
   a. How the information can be retrieved on web through various search engines.

b. How e-resources can meet their requirement in effective way

c. What are channels of searching the required information

5. After procurement of electronic resources, library professional should organize orientation programme for utmost use of e-resources.
6. Library professionals should also take the knowledge in depth to best and effective use of electronic resources.
7. Users should be motivated to use e-resources by personal meet.
8. Proper publicity of electronic resources, helps to make aware users about electronic resources

Conclusion

Libraries are to be the heart of any institutions, so Libraries should play their important role to take initiatives to familiarize library users about seamless access to e-resources instantly. Use of electronic resources, is associated with users awareness, So it is very important for library professionals to have keen interest to improve the awareness about e-resources and their uses, among users, especially scientific community, which leads to increased research productivity in agriculture sector. It is very necessary to have the proper and definite planning and programme to introduce and make aware about information resources to users by library and information professionals. There should be proper and specific guidelines to coordinate and manage users awareness of electronic information resources.

References

1. **Singh, Birajit A, Singh, Ibohal and Madhuri Devi;** *Users Skills and Awareness on Knowledge Retrieval Methods Adopted in Manipur Libraries : A Study*. Site : http://crl.du.ac.in/ical09/papers/index_files/ical-92_242_611_1_RV.pdf(visited on 12/3/2011).

2. **Mansuri, Imran**; *Using RSS Feeds and e-Alerts to Increase User Awareness of E-resources in Library and Information Centers.* Site : http://eprints.rclis.org/bitstream/10760/15319/1/final.pdf (visited on 17/3/2011).

3. **Salaam, M.O and Aderibigbe, NA**(2010); *Awareness and Utilization of The Essential Electronic Agricultural Library by Academic Staff* : A Case Study of University of Agriculture, Abeokuta, Nigeria. Chinese librarianship: an international electronic journal, 30. URL: http://www.iclc.us/cliej/cl30SA.pdf (visited on 18/3/2011).

4. **Tripathi, Harish and Hansraj**(2011); *Usage of Electronic Resources in ICAR Library.* National Conference Organized by IARI, held on 24-25 February, 2011.

5. **Kumbar, Mallinath, Rao,RV and Nargund, IN**; *Electronic Information Resources : A Methodological Approach Towards Introducing* users Site : http://shodhganga.inflibnet.ac.in/dxml/bitstream/handle/1944/111/cali_30.pdf?sequence=1 (visited 18/3/2011).

8

User Satisfaction in Digital Library Environment

Anil Kumar Dhiman &
Mahesh Singh Rana

Abstract

Digital libraries have emerged as an essentiality in modern environment. Any body can not think about to live without their existence. As user satisfaction is the ultimate aim of any library, so we have to devise some effective means and provide services to the users in digital era. This paper defines user satisfaction and various activities through which library can satisfy their users in digital environment.

Keywords: *Users satisfaction, Digital Libraries, Library Services.*

Introduction

The user community is the most important component of a library system. Every service in the library exists for the sole aim of satisfying its users and the indispensable goal of any library is user satisfaction. User satisfaction is an affective or cognitive state of mind which the user experiences as a result of his use and subsequent evaluation of library services.

User Satisfaction is an appraisal of perceived performance, which is a comparative process between various components, such as expectations and perceptions of services (Arishee, 2000). Hernon and Altman (1998) have identified two complicated perspectives for user satisfaction from the viewpoint of library service. First one is *service encounter satisfaction* – customer satisfaction or dissatisfaction with a specific service encounter; the other is *overall service satisfaction* – customer satisfaction or dissatisfaction with an organization based on multiple encounters or experiences. D'Elia and Walsh (1986) investigated user emotional satisfaction and have defined two types of satisfaction -"subjective user satisfaction," which corresponds to emotional satisfaction and "objective measures," which correspond to material satisfaction. Cullen (2001) has proposed the concepts of customer satisfaction at the micro-level and macro-levels to describe the complex interchange of customer expectations and perceptions across the services delivered by an organization. Concerning only one individual service, customer satisfaction at the micro-level contributes to the dimensions of service quality such as tangibles and reliability. Concerning all services with which the customer has interacted, customer satisfaction at the macro-level is a global view of the quality of service, and integrates many dimensions of service quality.

Service quality is defined as "a component of customer satisfaction and vice versa" (Cullen, 2001). Cullen further pointed out that it is unknown that when or in what circumstances, these concepts measure the same customer response and when or in what circumstances, they measure separate responses to service quality. Looking at these satisfaction concepts, Cullen drew a tentative relationship model as given in Fig. 1, to suggest the impact of user satisfactions on user loyalty.

Thus, it has become imperative in order to survive in an increasingly competitive information environment, and "...libraries must focus on improving the quality of the services they offer" (Simmonds and Andaleeb, 2001). So, library services have become more user-oriented and they are giving much attention on quality service to focus on customer satisfaction. Quality service can be delivered if libraries improve its management efficiency and the quality of library performance is inferred from the degree of satisfaction experienced by the users.

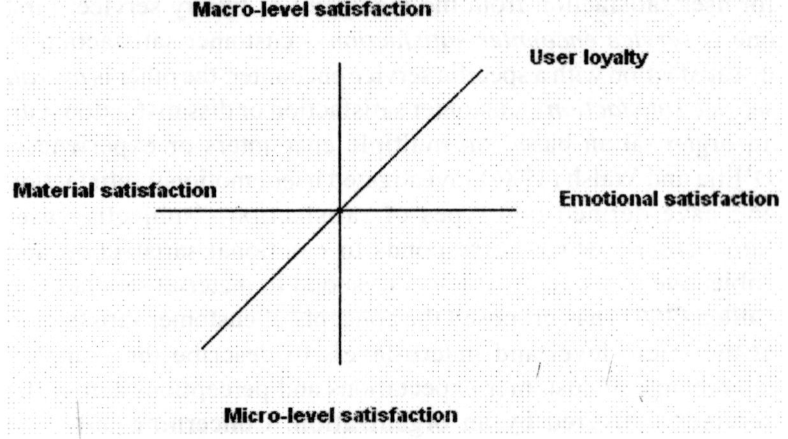

Fig. 1. Impact of Satisfaction on Customer Loyalty (After Cullen, 2001)

User Satisfaction in Digital Library Environment

Information and communication technology according to Mahapatra and Ramesh (2008) is the result of technological convergence of existing single isolated technologies viz., computer technology, communication technology, information processing, publishing technology etc. Digital Libraries have emerged as a result of such information and communication technologies development (Dhiman, 2003, Dhiman and Rani, 2011a). Digital libraries are the complex and advanced forms of information systems that can be endowed with a multiplicity of functions and features. They are the collections of information that have associated services delivered to user communities using a variety of technologies (Callan et al, 2000). However, Marchionini and Fox (1999) argue that digital libraries are extensions and augmentations of physical libraries. They suggest that with respect to this broader notion of place both physical and digital libraries are instantiations in different media of the same base type - the library. In either case, both physical and digital libraries occupy the physical-conceptual continuum with respect to ideas, materials, and people. The distinction between physical and digital libraries is thus not always a clear one. But the physical libraries are considered to maintain a collection of exclusively physical materials, while digital libraries are considered to maintain a collection of exclusively electronic materials. Thus, Sun Microsystems (2002) rightly defines a digital library as "the electronic extension of functions users typically perform and the resources they access in a traditional library".

Whatever the definition may be but the collections of information in digital libraries can be scientific, business or personal data and can be represented as a digital text, image, audio, video or other media including online web resources. It seems that due to the amount and great variety of information stored by digital libraries, they have become equal with the search engines in general- one of the major web services (Liaw and Huagn, 2003), which are faced by a diverse population of users who have heterogeneous background, skills, and preferences. Besides, the proliferation of digital libraries in the last few years has led to the development of various digital library collections and services. Every library and information service exists for its aim to satisfying its users, hence, a library should have to identify the specific community which needs its services as different users need not only different services, but also services at different levels to provide an efficient service (Seneviratne, 2008). This identification is essential for the systematic development of the collection and the planning of suitable services for the effective utilization of the collection in digital library environment for user satisfaction.

Riihelä et al (2009), have identified three types of user satisfaction as mentioned below:

Information Quality versus User Satisfaction

Information quality is the quality of the information produced by a system concerning its concentration on information utilization or consumption. Furthermore, it can be elaborated by features regarding the actual information that is presented by the information system. Information quality also encompasses variables that are mainly concerned with the user's information consumption such as flexibility, and the degree to which the available information is presented. User satisfaction can be obtained by transparency, management information and user-friendliness of catalogues in modern or digital libraries.

System Quality versus User Satisfaction

System Quality is recognized by technical features regarding the network and the information technology equipment itself. Therefore, some of the fundamental facets of system quality found in features like reliability, response time, and accuracy, ease of integration, flexibility, and functionality. So, system quality has to be assessed from multiple considerations where reliability is essential, making system integration an important aspect of them. The system quality variable can comprise of ease of use, system availability, and interoperability in digital library system. Here system quality concentrates on important characteristics or factors intrinsic to system design or implementation.

Service Quality Versus User Satisfaction

Service quality stands for certain aspects that service quality brings through information systems, which appear to fulfill the end-users demand in information technology era. This can be examined in terms of service consistency, reliability, timeliness, empathy, assurance, and accuracy or adequacy. Service quality variable can include responsiveness, accountability, compliance, process reliability and issue resolution, the help desk in digital environment.

Library Services for User Satisfaction

Since the users of today perform research in a fast, industrially oriented, innovative but also cost conscious manner, the topicality of library services and information is of overriding significance. The topicality of the library's services and information sources has become even more relevant with ever shorter half-lives, especially for scientific and technical economic information. Ball (2008) is of the opinion that library services for today's science sector must be precise, clear, unambiguous and relevant. The relevance of scientific information is extremely important, particularly in an era where we are inundated with information. It is the library's duty to ensure that the customer benefits from an information service that has been verified for relevance and is fit for use. Besides, the modern user of a library today expects complete information concerning his query. Only the extensive, complete and comprehensive processing of a user's information query using all available media and accessible sources is regarded as the fulfillment of their information needs. It is not enough to ensure the long-term existence of a library in order to meet quality criteria and achieve high customer satisfaction rather in negotiations with management, but concrete evidence must always be provided with regard to a library's monetary benefit, profitability, and how much it contributes to the success of the company as a whole. Satisfaction depends, to some extent, on patron expectations of services. It is appeared to be related to user perceptions of information accessibility, staff competence and helpfulness, computer usefulness and ease of use, and skill level for using libraries.

Thus, a library servicing its users must not only have current information at their disposal but also pass this information on to its customers rapidly and directly. The delays through any channels–due to the bureaucratic mentality, labyrinthine operations and high waiting piles may lead to customer dissatisfaction. A library must also be capable of making the information available to the customer in print or electronic form, by post, courier, and e-mails or as a coded universal resource locator. The following services and improvements based on Seneviratne (2008), and others, are to be implemented in library for user satisfied services in digital environment.

- A User guide with list of services offered should be prepared for distributing among the users in addition to the orientation sessions organized for new members.

- A presentation about the library services should be displayed on the computers.
- A web page, particularly of library with access to web catalogue should be crated and launched. External links and inlinks are to be provided in this web page for complete references of information.
- Users expect the staff to be informed about modern technology. So awareness programs should be organized to present information on various resources and services of the library for the use of its users.
- Seminar on web based information should also be carried out with collaboration of computer centre of the institute and or the library itself. This may increase the ratings about the web based information resources among the users.

Besides the proper updating signage and shelf labels and stack directories to reflect changes are also needed. Internet services for accessing various resources – whether they are open access or subscribed, should be provided to users. More and more emphasis should be given on collecting and developing e-resources in library collection because they provide quick access and retrieval of the information to its users in efficient ways. UGC-INFONET (Murthy et al., 2004) is one such library consortium which is providing access of e-resources to more than 146 universities throughout the country. INDEST–AICTE Consortium is also there which gives access to technical resources to IITs, IIMs and other AICTE approved institutions in the country (Arora and Tivedi, 2010). Both the consortia are providing access to e-resources country wide to a vast community of users and thus satisfying their needs.

Conclusion

Library is a place where enormous of information is acquired, stored, classified and disseminated to its users at the maximum to satisfy their needs. User satisfaction depends to a large extent on the ability of the library to integrate user needs into the development of the library. There are available various tools for customer retention like surveys which can be run on a regular basis. They help to detect weak points in library performance and give important hints for the optimization of library services. The participation of customers in the realization of open innovation projects guarantees immediate access to user needs and a high degree of user satisfaction in the long run. So emphasis should be given to provide as much library services as possible by the library to its users.

Management techniques like Total Quality Management (Dhiman, 1999; Dhiman and Rani, 2011b), Management by Objectives (Dhiman and Rani, 2004 & 2005) and Knowledge Management (Dhiman and Sharma, 2009) etc. can be used to judge the user needs and to train them in digital environment. Then we can improve the service accordingly to be remaining in the race of digital library development by satisfying the user needs and making them update.

References

1. **Arishee, J.** (2000). *Personal and Cultural Values as Factors in User Satisfaction: A Comparative Study of Users of Library Services*. Unpublished Doctoral Dissertation. University of Pittsburgh, Pittsburgh.
2. **Arora, Jagdish and Tivedi, Kruti.** (2010). INDEST–AICTE Consortium : Present Services and Future Endeavours. *DESIDOC Journal of Library & Information Technology,* 30 (2), 79-91.
3. **Ball, Rafael.** (2008). *User Satisfaction–The Key to a Library's Success*. Series III: e-Publications of the Institute ILS of the Jagiellonian University. Ed. Maria Kocójowa No 5. Library: The Key To Users' Success. 5-14. Available at -eprints.rclis.org/bitstream/10760/13994/1/ball-n.pdf.
4. **Callan, J., Smeaton, A., Beaulieu, M., Borlund, P., Brusilovsky, P. and Chalmers.** (2000). *Personalization and Recommender Systems in Digital Libraries*. Joint NSF-EU DELOS Working Group Report.
5. **Cullen, R.** (2001). *Perspectives on User Satisfaction Surveys*. Library Trends, 49 (4), 662-687.

6. **D'Elia, G. and Walsh, S.** (1986). Patrons' Uses and Evaluations of Library Services: A Comparison across Five Public Libraries. *Library & Information Science Research, 7,* 3-30.

7. **Dhiman, A.K.** (1999). *Implementation of TQM in Library and Information Centres.* In J. L. Sardana edited 'Library Vision 2000: Indian Library and Librarianship in Retrospects and Prospects (45th All India Library Conference Papers)'. Indian Library Association, New Delhi. Pp. 663 - 68.

8. **Dhiman, A.K.** (2003). *Basics of Information Technology for Librarians and Information Scientists.* 2 Vols. Ess Ess Publications, New Delhi.

9. **Dhiman, A.K. and Rani, Yashoda.** (2004). *Library Management: A Manual Book for Effective Management.* Ess Ess Publications, New Delhi.

10. **Dhiman, A.K. and Rani, Yashoda.** (2005). *Learn-Library Management.* Ess Ess Publications, New Delhi.

11. **Dhiman, A.K. and Sharma, H.** (2009). *Knowledge Management for Librarians.* Ess Ess Publications, New Delhi.

12. **Dhiman, A.K.and Rani, Yashoda.** (2011a). *Digital Libraries.* Ess Ess Publications, New Delhi.

13. **Dhiman, A.K. and Rani, Yashoda.** (2011b). *Total Quality Management for Libraries in Digital Environment.* In Manowwer Eqbal edited 'Information Professional: Issues and Challenges in Digital Age.' Pragun Publications, New Delhi. Pp. 365-72.

14. *Hernon P. and Altman, E.* (1998). *Assessing Service Quality: Satisfying the Expectations of Library Customers.* American Library Association, Chicago.

15. **Liaw, S. and Huang, H.** (2003). An Investigation of User Attitude toward Search Engines as an Information Retrieval Tool. *Computers in Human Behavior,* 19 (6), 751-765.

16. **Mahapatra, M. and Ramesh, D.B.** (2008). *Information Technology Application in Libraries- A Text Book for Beginners.* Bhubneshwar.

17. **Marchionini, G. and Fox, E.** (1999). Editorial: Progress toward Digital Libraries: Augmentation through Integration. *Information Processing & Management,* 35 (3), 219-225.

18. **Murthy, T.A.V. et al.** (2004). Access to Scholarly Journals and Databases: UGC-INFONET E-Journals Consortium. *University News.* 42 (34) : 1-5, 8.

19. **Riihelä, Timo, Mattila, Minna and Haapaniemi, Harri.** (Eds.). (2009). *Beyond the Dawn of Innovation 2009 - Conference Proceedings.* Laurea University of Applied Sciences, 15 – 17 June, Finland, Vantaa.

20. **Seneviratne, Daisy.** (2008). Measuring User Satisfaction: A Case Study at the PGIM Branch Library at Peradeniya. *Journal of the University Librarians Association of Sri Lanka,* 10, 40-53.

21. **Simmonds, P. L. and Andaleeb, S. S.** (2001). Usage of Academic Libraries: the Role of Service Quality, Resources and User Characteristics. *Library Trends,* 49 (4), 626-634.

22. **Sun Microsystems.** (2002). *Digital Library Trends.* Available at http://www.sun.com /productsn-solutions/edu/whitepapers/pdf/ digital_library_trends.pdf.

9

Consortium Based Subscription of Electronic Resources : A Case Study of Various Indian Library Consortium

M. Ananda Murugan

Abstract

This paper traces the history of Indian library consortium and its variety of subscriptions for the academic people of Engineering technology, management, Agricultural, and Medicine. For provide equal access of electronic resource with low cost methods these consortiums are playing role for selecting the electronic resource based on the research and academic needs of various type of universities like technical , medical , agricultural and Management and also approach the Government for getting fund. This article provides the information about the various consortium and its objectives, function of the councils, members of the consortium, and details of the electronic resources.

Introduction

Over the past decade number of library consortia operating across different type of libraries and across different geographical boundaries. In India many consortium formed based on the group of Institution like Management, Medicine, Agricultural science and other academic Universities. The aim of the consortium is to provide a common interface with familiar functionality for the benefit of library patron and the Indian research community. And maximize the value of the intellectual investment involved in selecting and implementing the shared system and in creating and operating a management structure for the benefit of the member of the consortium. But most of the consortiums are needed to develop according to the user community of India. In this article describe the over view of Indian consortium, its structure and resource available through the consortium to the members.

INDEST (Indian National Digital Library in Engineering Sciences and Technology Consortium)

In India, after Independence many Engineering Instructions are started and now it has flourished and become a Campus of producing talented manpower's in the world intellectual market. The government started many new IITs based on the need of manpower in India and also world. Many new engineering colleges are allowed to start by Private people to provide better education for the rural people in India. And the Indian government spends crores of rupees for subscription of Print and online journals to the students and faculties of these institutes. To avoid duplicate of purchase the same materials and procure more materials in the form of E-Journals, Databases and give the uniform access of all resources for all users

The Ministry of Human Resource Development (MHRD) has set-up the "Indian National Digital Library in Engineering Sciences and Technology (INDEST) Consortium" on the recommendation made by the Expert Group appointed by the ministry under the chairmanship of Prof. N. Balakrishnan. The Ministry provides funds required for subscription to electronic resources for (48) institutions including IISc, IITs, NITs, IIMs and a few other centrally-funded Government institutions through the consortium headquarters set-up at the IIT Delhi. Besides, (60) Government or

Government-aided engineering colleges and technical departments in universities have joined the Consortium with financial support from the AICTE. Moreover, the INDEST-AICTE Consortium, as an open-ended proposition, welcomes other institutions to join it on their own for sharing benefits it offers in terms of highly discounted rates of subscription and better terms of agreement with the publishers. All electronic resources being subscribed are available from the publisher's Website. The Consortium has an active mailing list and a Web site hosted at the IIT Delhi.

The INDEST-AICTE Consortium is the most ambitious initiative taken so far in the country. The benefit of consortia-based subscription to electronic resources is not confined to 38 major technological institutions in the country but is also extended to all AICTE-accredited and UGC-affiliated institutions. (1096) engineering colleges and institutions have already joined the consortium on their own. Recently (198) engineering colleges and institutions joined under self support- new scheme .

The members of core group of Institutions are as follows:

1. IITs and IISc (8)
2. NITs , ISM, SLIET and NERIST (21)
3. IISERs (5)
4. IIMs, IIIT, IIITM and NITIE (14)
5. AICTE - supported Members (60)

   AICTE provides funds for access to e-resources to (60)Government engineering colleges or technical institutions.

6. Self-supported Institutions (1096)

   The consortium, being an open-ended proposition, invites AICTE-accredited and UGC-affiliated institutions to join hands with the leading Engineering and Technological Institutions in India and share the benefits it offers in terms of lower subscription rates and better terms of agreement with the publishers. (1096) other engineering colleges and institutions have already joined the consortium on their own.

7. Self- supported members under new scheme (198)

UGC-INFLIBNET

India possesses a highly developed higher education system which offers facility of education and training in almost all aspects of human creative and intellectual endeavors: arts and humanities; natural, mathematical and social sciences, engineering; medicine; dentistry; agriculture; education; law; commerce and management; music and performing arts; national and foreign languages; culture; communications etc. The institutional framework consists of Universities established by an Act of Parliament (Central Universities) or of a State Legislaturc (State Universities), Deemed Universities (institutions which have been accorded the status of a university with authority to award their own degrees through central government notification), Institutes of National Importance (prestigious institutions awarded the said status by Parliament), Institutions established State Legislative Act and colleges affiliated to the University (both government-aided and –unaided) As on 31.3.2010, there were 367 University level institutions including 40 Central Universities, 251 State Universities, 130 Deemed Universities and 5 institutions established under State Legislation, 13 Institutes of National Importance established under Central legislation and 6 Private Universities.

Due to increase of Universities and its research activates the Government decided to develop the unique center for supply of information resource. Based on the view UGC INflibnet was started in India on 1996.

Information and Library Network (INFLIBNET) Centre is an autonomous Inter-University Centre of the University Grants Commission (UGC) of India. It is a major National Programme initiated by the UGC in 1991 with its Head Quarters at Gujarat University Campus, Ahmedabad. Initially started as a project under the IUCAA, it became an independent Inter-University Centre in 1996.

INFLIBNET is involved in modernizing university libraries in India and connecting them as well as information centres in the country through a nation-wide high speed data network using the state-of-art technologies for the optimum utilisation of information. INFLIBNET is set out to be a major player in promoting scholarly communication among academicians and researchers in India.

Member of the Consortium:-

1. Universities covered under PhaseI (50)
2. Universities covered under PhaseII (50)
3. Universities covered under PhaseIII (76)
4. Associate Members (97)
5. IUCs and Other Institutions (5)
6. National Law Scools/Universities (14)

IIMs Library Consortium

Indian Institute of Management Libraries are well equipped with Management resources in the form of technical books, Journals and online databases. In 2000 all the IIMs started to form a consortium for the resource and save the government money from the duplication of many resources. IIMK took leadership in the formation of the IIM Consortia, and had the first meeting in Calicut. The objective was to ensure among the IIMs, optimum utilization and enhancement of the resources, and to minimize the expenditure by consortia based subscriptions to the commonly subscribed databases and journals. The idea was to approach publishers of CD-ROM Databases to begin with, as a consortium, for better pricing and services. Eventually, other digital databases and journals were also planned to be covered by the programme. The meetings were proved to be very productive and successful. During the first meeting itself, a host of databases were jointly purchased at very competitive prices, and a number of others promised supply of their products at a nominal cost.

In the case of journals, all the six IIMs put together subscribes to over 2550 scholarly titles of which around 1200 are duplications (overlapping titles). Among these, 33 titles are being subscribed to by all the IIMs. Having convinced on the dire need for a journals consortia, major publishers such as Elsevier, Kluwer, Wiley, Blackwell and MCB University Press were approached and they all represented in the second meet which was held at IIM Bangalore in 2001. The end result has been highly praiseworthy, that over 740 E-journals IIMs are able to get online access, across all the IIMs, by paying a nominal additional amount..

Medical Library Consortium

NML, National Medical library consortium started in the year 2008 with 40 members including 28 ICMR (Indian council of Medical research), 10 DGHS (Directorate General of Health service), AIIMS, NTR Health University. The first initiative taken by Director General of Health Service to develop Electronic Resources in Medicine (ERMED) Consortium is an initiative taken by Director General of Heath Services (DGHS) to develop nation wide electronic information resources in the field of medicine for delivering effective health care for all. The authorities decided to provide financial support required for the purchase of electronic journals under the ERMED consortium project for Government medical colleges and institutions. For private Medical colleges/ institutions the payment per site for e-resources purchased by the consortium will be charged. The Consortium is being coordinated through it's headquarter set up at the National Medical Library (NML).

NML Started ERMED e-journal consortium in 2008 with 40 members including 28 ICMR+10 DGHS + AIIMS + NTR Health University, Vijaywada, Andhra Pradesh. It provided electronic journals from 5 publishers, which contained 1515 medical journals. The ERMED invested Rs. 2.6 Crore for electronic journals in 2008. The ERMED resources can be searched and browsed through a versatile search platform www.nmlermed.in.

In 2009 the ERMED members increased from 40 to 72 Government Medical Colleges/Institutes across the country. NML paid Rs. 7.55 crore for Government Medical Colleges/Institutes. The letter for self payment has been sent to some Private Medical

Colleges/Institutes. ERMED is providing over 1600 medical journals in 2009 from 9 leading medical journal publishers.

Aim & Benefits of e-journal Consortium

- Round-the clock instant online access to multiple users through IP address and customer ID.
- Access to users beyond the physical space and time of the library. Users can access library's e-journal resources from their Departments.
- Consortia models offered by publishers may help in:
  - o Benefit of cross sharing
  - o Resource increase by depth (back volume) and breadth (non subscribed title) to consortium members.
  - o Negotiable price for subscribed titles.
- Ultimate aim of any e-journal consortium is to make online journal literature available to unreachable medical scholars working in the country through electronic media.
- Facilitates better management of information resources in electronic environment.
- Hassles of archiving print resources and their management is reduced.
- Dissemination of e-journal literature is more fast, more economical and efficient.
- NML is already providing over 3380 print journal articles/ month across the country through postal service. Efficiency of the same may be enhanced if electronic options are available to NML.

Agricultural Library Consortium

Consortium for e-Resources in Agriculture : Agricultural research, the backbone of agricultural growth in the country, demands timely dissemination of knowledge being generated and updated across the globe from time to time. With the advent of internet facilities and advancement of web technology, almost all reputed international journals are available on-line and can easily be accessed by researchers over the network. Since ICAR is having network connectivity across the institutes and state agricultural universities, select journals could be made available over the network for the use of scientific community. Keeping this broad objective in mind, the National Agricultural Innovation Project (NAIP) has funded for establishing the Consortium for e-Resources in Agriculture (CeRA) at the Indian Agricultural Research Institute (IARI).

Objectives of CeRA

- To develop the existing R & D information resource base of ICAR Institutes/universities, etc., comparable to that existing in world's leading institutions/organizations.
- To subscribe e-journals and create an e-access culture among Scientists/teachers in ICAR institutes/agricultural universities.
- To develop a Science Citation Index (SCI) Facility at IARI for evaluation of scientific publications.
- To assess the impact of CeRA on the level of research publications measured through SCI.

Work Plan

The work plan to meet the above objectives comprises

- The establishment of a CeRA Co-ordination cell for negotiation and Subscription of e-journals.
- Formation of mandatory Committees for guiding and monitoring the work of the Consortium.
- Installation and maintenance of dedicated server and e-storage facility for maintaining archival repository of e-resources in electronic form.
- Development/adoption of suitable software for e-access.
- Identification of e-journals/publishers/vendors and procurement of e-journals
- Organization of workshops/trainings to facilitate and popularize e-access and

- Monitoring and impact analysis of e-resources through web of science and NASS ID.

The Consortium consists of the following organizations under NARS:

1. Deemed Universities
2. ICAR Headquarters and Krishi Anusandhan Bhawans
3. National Bureau
4. Institutes
5. National Research Centers
6. Project Directorates
7. State Agricultural Universities

Resources Available at Major Consortiums

INDEST Consortium	*UGC INFONET Consortium*	*National Medical Library Consortium*
ABI / Inform Complete	American Chemical Society	American Academy of Pediatrics
ACM Digital Library	American Institute of Physics	BMJ Publishing
ASCE Journals	American Physical Society	Cambridge University Press
ASME Journals	Wiley-Blackwell Publishing	Cengage Learning
ASTM Standards & Digital Library	Cambridge University Press	IOS Press
Capitaline	Elsevier Science	Lippincott William Wilkins
CRIS INFAC Ind. Information	Emerald	Oxford University Press
McGraw-Hill's Access Engineering (FKA DEL)	Economic & Political Weekly (EPW)	Proquest
EBSCO Databases	Hein Online	Royal Society of Medicine Press
Elsevier's Science Direct	Institute of Physics	John Wiley and Sons

Contd....

INDEST Consortium	*UGC INFONET Consortium*	*National Medical Library Consortium*
Emerald Full-text	J-STOR	Informa Healthcare
Emerald Management Xtra	Manupatra	Chinese Pharmacological Society
Engineering Science Data Unit (ESDU)	Nature	Scandinavian Physiological Society
Euromonitor (GMID)	Oxford University Press	British Medical Acupuncture Society
IEEE / IEE Electronic Library Online (IEL)	Portland Press	Sage Publications
Indian Standards	Project Euclid	Oxford University Press
INSIGHT	Project Muse	American College of Physicians
Nature	Royal Society of Chemistry	British Medical Acupuncture Society
ProQuest Science (formerly ASTP)	SIAM	Guilford Press
Springer Link	Springer Link	Thomson American Health Consultants
IET Digital Library	Taylor and Francis	National Institute on Alcohol Abuse and Alcoholism
Emerald E-books	Westlaw India	Royal Swedish Academy of Sciences
INFORMS Pub Suite	SciFinder Scholar	American Society for Nutrition
GALE Gale Cengage Learning, (IEC, BCRC and GREENR	MathSciNet	American Society for Clinical Pathology
ICE Publishing (Thomas Telford) Journals	Royal Society of Chemistry (6 Databases)	American Association of Critical-Care Nurses
SCOPUS Database	ISID	International Society of Dermatopathology
INSPEC on EI Village	Web of Science (Through N-LIST Programme)	Science Publications

Contd....

INDEST Consortium	*UGC INFONET Consortium*	*National Medical Library Consortium*
MathSciNet		American Nurses Association
SciFinder Scholar		American Occupational Therapy Association
Web of Science		American Speech-Language-Hearing Association.etc.
Optical Society of America		
Project Muse		

Need to Develop the Consortiums as Like in Foreign Countries

In India the consortium are working look like the suppliers or agent for subscribing the online resource like e journals, e-books and other bibliography databases. But the concept of consortium is not only that and it should be as follows :

1. Consortium should provide a common interface with familiar functionality for the benefit of library patrons.
2. Maximize the value of the intellectual investment involved in selecting and implementing the shared system and in creating and operating a management structure for the benefits of the members of the consortium.
3. Provide opportunities for shared problem solving amongst Library professionals.
4. Members of the consortium have a sharing infrastructure includes economic benefits.
5. The consortium has to maintain the union catalogue of books and journals.
6. The consortium has responsible to solve the technical difficulties of the library during the implementation of new technology and Uniform Computer Information Transactions.
7. The consortium should be the whole responsible of maintain the uniformity of the member library.

Conclusion

The role of library consortium is to support the member libraries in all aspect like sharing of information, ideas, and expertise. In Indian almost all type of Libraries formed the consortium and providing the service to the member libraries. But the value of the consortium should be Respect for all libraries and librarians, Responsiveness to member needs, Equity of service across types and sizes of libraries. Most of the Indian library consortiums are concentrating on the purchase of on line resource for its member libraries. They have to analysis the status of member libraries and provide maximum help to develop the library for better usage of the resource. The Agricultural library consortium and Management Library consortiums are needed to more function like UGC Library consortium.

References

1. **Nfila, R.B & Darko-Amphen.** (2002). The development in academic library consortia from the 1960's through to 2000: a review of literature, *Library Management,* 23(4): 203-212.
2. **Rao, Siriginidi S.** (2001).Networking of libraries and information centers: challenges in India, *Library Hi Tech,* 19(2).
3. **Rona, Wade.** (1999). The very model of a modern library consortium, *Library consortium management: An international journal* 1(1): 5-18.
3. **Scepanski, Jordan M.** (1998). Collaborating on new missions-library consortia and the Future of academic libraries. *Conference on New Missions of Academic Libraries in the 21stCentury,* Beijing: Peking, 25-30 October. Electronic address: www.ait.org.tw/events/docs/20070501-Jordan.pdf.
4. **Wikipedia** (2007). *Library Consortia in the world,* Electronic address : htt://np.wikipedia.org.
5. UGC-Infonet, http://www.inflibnet.ac.in/infonet/
6. INDEST-AICTE Consortium, http://paniit.iitd.ac.in/indest/

10

Role of E-Books in Libraries : A New Horizon for us

Amrita Mjaumdar

Abstract

With growth and development of web technology the popularity of e-books increased and the academics are actively building electronic information resources for providing value added services to users to meet their information needs. Present paper briefly discusses about the e-book with role significance and challenges in arena of academic library.In addition to that it also discuses the future compatibility with the printed books.

Introduction

The journey of e book stars from 1971 with project Gutenberg until now with rapid growth of web technology Amazon, Adobe, Mobipocket, Google Books, the Internet Archive, and many others. Make their imprint in publishing electronic book. The production and usage of e-books is growing at a fast pace and warrants serious consideration, especially in libraries and information centers of Indian universities, institutions of Higher Education, S & T Organizations and other institutions of national importance like IITs, NITs, IIMs, and IISc. They are actively engaged in building electronic information resources for providing value added services to users to meet their information needs. The present paper discuses on eBook with its type of vendor and their role . It also discussed the advantage and limitations of e-books in different environment like bandwidth, connectivity etc. Lastly, it focuses on future of e-book in respect of printed book that formulate a workable and sustainable strategy in academic Libraries. It also urgently needed that systematic efforts are taken to develop suitable information infrastructure for providing access to e-books.

Definition

The concepts of the "electronic book" are varied widely. Some attempts have been made to include different definitions: The **Webopedia** defines as: "An electronic version of a book. Currently there are two e-book products available, the Rocket eBook, from Nuvomedia (www.nuvomedia.com) and the SoftBook from SoftBook Press (www.softbook.com). Both are small computers size of a paperback and a legal notepad with back lighted screens that allow a user to read, save, highlight, bookmark, and annotate text. Both can download books from a Web site, such as barnesandnoble.com (although the Rocket e-Book requires another PC).

According to **Ana Arias Terry** "an e-book consists of electronic content originating from traditional books, reference material, or magazines that is downloaded from the Internet and viewed through any number of hardware devices." These include PCs, laptops, PDA's (personal digital assistants), palm PC's or palmtops, or dedicated e-book readers.

Therefore, the definition can be integrated as "An electronic book is the electronic content that is transmitted and or displayed on a device or system to be read by the viewer similar in experience to reading a physical book. An e-book is written in machine-readable form (Computer Screen) down loaded to a PC of digital assistant or place on a reader design for that purpose. It may be referred to a document of written record or electronic text. It can be delivered by down loading or as e-mail file attachment.

Here is the "virtual" journey we are going to follow:

- 1971: Project Gutenberg is the first digital library
- 1990: The web boosts the internet
- 1993: The Online Books Page is a list of free e-books
- 1994: Some publishers get bold and go digital
- 1995: Amazon.com is the first main online bookstore
- 1996: There are more and more texts online
- 1997: Multimedia convergence and employment
- 1998: Libraries take over the web
- 1999: Librarians get digital
- 2000: Information is available in many languages
- 2001: Copyright, copyleft and Creative Commons
- 2002: A web of knowledge
- 2003: eBooks are sold worldwide
- 2004: Authors are creative on the net
- 2005: Google gets interested in e-books
- 2006: Towards a world public digital library
- 2007: We read on various electronic devices
- 2008: "A common information space in which we communicate.

Significance of E Books

It is observed that e-books have great potential and bright future to attract users. In addition to this fact, there are some other major factors which also motivate academics to consider developing e-books are as follows:

1. Technical significance
   - Can be updated and, stored very easily
   - Can be downloaded instantly
2. Economic
   - Users can read an e-book any time
   - Contain the latest and most updated information

- In buying e-books, the overhead charges like shipping, postal, handling are totally ruled out
- No cost of technical processing and maintenance
- Link can be created in the record of OPAC
- No risk of book theft and tearing and mutilation of pages
- E-books save library space
- E-books do not require bindery and repair
- E-books save human resources for shelving and rectification
- Users can not misplace e-books

3. User friendly
   - Provide facility to hold and turn pages easily
   - Physically disabled users can hear audible e-book
   - Due to portability, e-Books can be taken any where on portable computer
   - Font size can be changed suitably.
   - Provide facility to hold and turn pages easily
   - E-books have background music and animations.
   - Hyper linking enables users to communicate directly with author

Types of E-Books

There are different types of e-books available in the market. Some of the types are explained in this paper:

Downloadable e-Book The contents of e-book are available on a website for downloading to the user's system. The users do not have to purchase any special reading device and can employ standard and well-known web techniques to obtain the book. The disadvantages of downloading e-book include the problems of reading from PC screens, unattractive formats, and the time required to carry out downloading operation, particularly in the absence of high speed data lines. Many consumers have only 56K nodes and dial up connections that can be slow, overloaded and unreliable.

Dedicated e Book

The contents of the books are downloadable to a dedicated hardware device, which has a high quality screen and a special capability for book reading. Much of the activities of dedicated e-book arena centers around the emergence of dedicated e-book readers-hardware devices specially built and designed to improve the reading experience and they incorporate special control to make book reading easy and simple. They have also the facilities for book marking a page, move through the book in a nonlinear fashion. Some readers also incorporate links to dictionaries or thesaurus so that the user can look up the meaning of the words. There is no need for PC or Internet access facilities, because the readers incorporate modems that dial directly into the e-book publisher's server to download books.

Web Accessible e-Books

The book remains on the providers' web site and can be accessed on a fee basis. Readers can purchase the books to receive indefinite access. Users require PC to access this kind of e-book.

Print-on-Demand Book

The content of a book is stored in a system connected to a high speed, high quality printer from which printed and bound copies are produced on demand. The contents may be accessible chapter by chapter basis, to enable the creation of single copies of customized books.

E-Book Providers and Accessibility

Electronic books are accessible via the Web in a number of forms. Generally, they are texts that have been scanned or typed and either published on a Web server or made available on the Web for downloading. These can be free access or access on payment.

FREE Accessibility Project Gutenberg (http://www.promo.net/pg)

Project Gutenberg is the Internet's oldest producer of FREE electronic books (e-Books or e-Texts). It is the brainchild of Michael Hart in 1971. Thousand of e-books are available in this site free of charge. The site is having browsing facility by Author , and by Title. It also provides advanced search facility by Author or by Title (Words). Author wise list of e-books is available at http://www.promo.net/pg/authors.zip and Title wise list of e-books is available at http://promo.net/pg/titles.zip and current list of e-books is available as GUTINDEXALL at the FTP site at the University of North Carolina.

Bartleby.com (http://www.bartleby.com)

It is a comprehensive, searchable database of reference, verse and classic literature, summaries of each book, and author biographies, having enhanced navigational tools and extensive cross–referencing. No access fee is charged. The screen shot of the web allowing access to e-books to the users is given below:

University of Virginia e-book Library (http://etext.lib.virginia.edu/ebooks/ebooklist.html

1,800 publicly-available e-books including classic British and American fiction, major authors, children's literature, the Bible, Shakespeare, American history, African-American documents, and much more are available. Thousands more titles can be found on electronic text center.

University of Illinois Electronic books (http://www.press.uillinois.edu/epub/books.html)

The University of Illinois Press is a growing archive of free e-books, most with full content, features electronic versions of front list and newly released titles. It provides to browse through this collection by author, title and subject free of charge. It also provide full keyword searching across the archive to help the user to locate specific information.

University of Pennsylvania Online Books (http://onlinebooks.library.upenn.edu/)

The Pennsylvania Online Book page facilitates access to books that are freely available over the Internet. This site includes different sections like Books Online, News, Features, Archives. It was founded and is edited by John Mark Ockerbloom, the digital library planner and researcher at the University of Pennsylvania. It includes over 20,000 free books on the Web in various formats. Recently it is updated on October 7, 2004. All are free for personal, noncommercial use. Users can search by author and title, browse new listing, by author, by title, by subject and can browse serial archives.The screen shot of the web allowing access to e-books to the users is given below:

The National Academies Press (http://www.nap.edu/)

The National Academies Press (NAP) was created by the National Academics to publish the reports issued by the National Academy of sciences, the National Academy of Engineering, the Institute of Medicine and the National Research Council, operating under a charter granted by the Congress of the United States. NAP publishes over 200 books a year on a wide range of topics in science, engineering, and health, capturing the most authoritative views on important issues in science and health policy. The institutions represented by NAP are unique in that they attract the nation's leading experts in every field to serve on their blue ribbon panels and committees. More than 3,000 books online Free and more than 900 PDFs now available for sale.

Web Book Publication (http://www.web-Books.com)

All titles in this public domain are free for online reading and download in EXE and/or ZIP formats (collectively called Web–books. The EXE format can be read by Internet Explorer on Windows 95/98/XP. The ZIP format is a collection of HTML files that can be read by any web browser on any device (notebook, PDA, smartphone, etc.). Currently it has over 700 titles. More titles will be added regularly

Digital book Index (http://www.digitalbookindex.com)

This site is expanded to 91,000 e-books and out of 91,000 e-books more than 50,000 e-books are freely available. Before accessing this e-book database the users have to identify themselves for better understanding of the visitors and to identify the kinds of resources the visitors are looking for. They also make user e-mail addresses available to partner publishers who want to offer their new and current titles to the visitors.

netLibrary (http://www.netlibrary.com/Gateway.aspx)

netLibrary is founded in 1999 and based in Boulder, Calorado, U.S.A. and become a division of OCLC on January 25, 2002. It currently provides about 60,000 e-books from 400 publishers and adds 2,000 additional titles per month. Full list of books is available at URL: http://www.netlibrary.com/titleselect. The user can create his/her own login and password before access this page. The list of publishers of netLibrary is also available at URL: http://www.netlibrary.com/about_us/publishers/publisher_list.asp netLibrary e-books can also be linked to library's website or integrated with existing cataloguing systems and OPAC using referring URLs behind a password protected page or an authentication system.

Kluwer Online (http://www.kluweronline.com/ebooks/sales/)

(Recently merged with Springer)

Kluwer Online's new eBook platform makes the accessibility of its online eBook collection easier and more convenient for subscribing institutions. IP-enabled authentication is all that's necessary to purchase and view Kluwer's comprehensive array of electronic research information. A subscription to Kluwer Online eBooks provides:

- Over 650 comprehensive, current works in the scientific, technical, and medical fields
- Online viewing through Adobe® Reader®—no additional software to install or large files to download
- Full-text searching across the entire catalogue

- Usage statistics provided monthly in spreadsheet format
- No physical inventory, no shelving, and no lost or overdue books to be concerned about.
- Archive titles, control pricing, track usage and maximize the budget.

Other E-Books Providers

In addition to netLibrary, Kluwer online etc. dealing with engineering and technology subjects there are numbers of e-books providers in different variety of subjects like fiction, history, entertainment, college books. Some of the e-book providers dealing with such subjects are stated below:

Questia (http://www.questia.com)

Questia is the world's largest online collection of books and journal articles in the humanities and social sciences, plus magazine and newspaper articles. The users can search each and every word of all of the books and journal articles in the collection. They can read every title cover to cover. This rich, scholarly content selected by professional collection development librarians is not available elsewhere on the Internet. Undergraduate, high school, graduate students, and Internet users of all ages have found Questia to be an invaluable online resource. Anyone doing research or just interested in topics that touch on the humanities and social sciences will find titles of interest in Questia.

To complement the library, Questia offers a range of search, note-taking, and writing tools. These tools help students locate the most relevant information on their topics quickly, quote and cite correctly, and create properly formatted footnotes and bibliographies automatically. Questia provides a comprehensive research environment to meet students' academic needs.

E-Book Palace (http://www.ebookpalace.com/)

eBook Palace is a visitor submitted directory listing over 3,500 titles in popular ebook formats including pdf, lit, html, exe, palm, mobipocket, & Rocket Editions that users can download and read from their desktop, handheld device or read online.

BarnesandNoble.com (http://www.barnesandnoble.com/)

It carries downloadable e-books in their E-Book section. CD ROM and diskette e-books available in the general book section.

Booklocker.com (http://www.booklocker.com)

Booklocker.com, Inc. develops low-cost, author-friendly products, services and programs that help authors publish and market their own works. Their offerings work best with entrepreneurial authors who actively market their books and want to be involved in the entire process. Their goal is to provide the tools and the knowledge to help them successfully sell their books.

Book-Studio/Studio E(http://www.STUDIO-E-BOOKS.COM)

Book-Studio is a complete book production service. From raw manuscript to delivered book, Studio E provides editing, text preparation, book design, typesetting, and digital or camera-ready layouts ready for the printer. They are also broker of printing and shipping, delivering finished books ready to read.

Word Wrangler (http://www.wordwrangler.com)

Word Wrangler Publishing , 332 Tobin Creek Road, Livingston, MT 59047 is publishing fiction and non-fiction E-books partnering with SynergEbooks.

Palm Digital Media (http://www.peanutpress.com)

Palm Digital Media is the first electronic book publisher to offer contemporary fiction and non-fiction books, newspapers, and magazines for reading on handheld computers including the Palm Organizer, Sony Clié, Handspring Visor, and Pocket PC machines, as well as Windows and Macintosh computers.

Powells.com (http://www.powells.com)

This is largest new and used bookstore, carries a large selection of Rocket Editions contains rare books , technical books and schooling books.

Limitation of E-Book

Though the e-book has great potential and bright future to attract the users it also have some limitations.

- Physical and Mental Strain: to read an e-book is not comfortable in comparison with printed book. It cannot be read in leisure time lying on bed or sofa. It can't be read in open air. Continuous reading of an e-book sitting in front of computer system affects the eyes and nerves.
- Reading an e-book while traveling in train or bus is not so easy and cheaper in comparison with printed books. Laptop or Palm top with battery system is urgently necessary to read an e-book, which is more costly than a book.
- In addition to cost, the user has to carry the machine which is difficult and a matter of risk like break / damage or theft.
- Students, faculties and research scholars those are staying in campus may avail the computer facility during working hours but if any student stays out side campus has to purchase computer and Internet connectivity for reading online e-books. Common students cannot afford such financial burden. Only in campus students, faculties and research scholars of IIT's, IISc., and NIT's/ REC's can avail this facilities. In India there are plenty of private Technical & Engineering Collages who can't afford even in campus facilities.
- E-book is costlier than printed book. The expenditure of an e-book is sum total cost of e-book plus access charges plus internet connectivity and usage charges which is costlier than that of a printed book.
- In addition to Indian Scenario the survey of Task Force of the University of California 2001 found that most institutions are still in trial stage with e-books and e-book market viable are not quite in pace.

Perspective and Future of E-Book

The e-books are having so many disadvantages and problems but still e-book has its perspective due to the following features:

- Malleability
- Flexibility
- Search ability
- Ubiquity

As per statistics available from open e-book forum June 22, 2004 the sales and production of book are gradually increasing. A comparative statement of sales, production and reference is stated in the following figure.

1. First half 2003——$288,440
2. First half 2004——$421,955

Libraries are still defined by the printed book. Most academic librarians that we know are so much more than just a printed book. But in the subject like management, Science& technology requires currency at a level so this will inclined towards e books Promoters and publishers of traditional books often think that ebooks are in danger of replacing paper books. It's not likely that ebooks will ever be able to completely and totally replace print books, nor should they. Regardless of future developments in e-reading technology, the book market will have plenty of room for paper books for quite some time. However, the print book market will change and shift as a result of the ebook.

In the present, and for a large chunk of the foreseeable future, paper books will continue to provide the primary source for textual information, and traditional books will continue to be a commonplace in our culture. It seems that ebooks and print books are able to complement each other's strengths and weaknesses quite well.

References

1. **Gupta K.K. and Gupta P.K.** *E-Books : A new media for Libraries.* Paper presented at CALIBER-2002 held at Jaipur, February, 14-16, 2002. pp. 384-399.
2. **Jala Samir Kumar.** *Electronic Book: a kind of digital resource.* Paper presented at CALIBER-2001 held at Pune University, Pune, March, 15-16, 2001. pp.24-31.
3. **Harish Chandra.** *Development and Use of Web Based Information Resources with Specific Reference to e-Books : A case for S&T Libraries.* Paper presented at International CALIBER-2003 held at NERF, Ahmedabad, February 13-15, 2003. pp. 575-586.
4. Internet.com (*Webopedia)*. http://webopedia.internet.com/TERM/e/electronic_book.html. Last visited 10/07/2004.
5. **Landoni M. and Gibb F.** The role of visual rhetoric in the design and production of electronic books: the visual book. *The electronic library.* 18 (3). 2000.
6. **Ploama Diaz.** *Usability of Hypermedia Educational e-Books.* D-Lib Magazine, 9 (3), March, 2003 available at: http://www.dlib.org/dlib/march03/diaz/03diaz.html. Last visited 05/07/2004
7. **Snowhill Lucia.** *E-books and Their Future in Academic Libraries.* D-Lib magazine, July/August, 2001 available at: http://www.dlib.org/dlib/july01/snowhill/07snowhill.html Last visitd 05/07.2004 TERRY (ANA ARIAS). Demystifying the e-Book - what is it, where will it lead us, and who's in the game? *Against the grain.* November 1999. Available: http://bibliofuture.homepage.com/demyst.htm.
8. http://etext.lib.virginia.edu/ebooks/ebooklist.html
9. http://proquest.safaribooksonline.com
10. http://www.bartleby.com
11. http://www.booklocker.com
12. http://www.cyberread.com
13. http://www.digitalbookindex.com
14. http://www.ebookpalace.com/
15. http://www.kluweronline.com/ebooks/sales/

16. http://www.nap.edu/
17. http://www.netlibrary.com/Gateway.aspx
18. http://www.openebook.org/pressroom/pressreleases/q104stats.htm
19. http://www.peanutpress.com
20. http://www.powells.com
21. http://www.press.uillinois.edu/epub/books.html
22. http://www.promo.net/pg
23. http://www.questia.com
24. http://www.studio-e-books.com
25. http://www.web-Books.com
26. http://www.wiley.com
27. http://www.wordwrangler.com

11

E-Resources : Ease in Access

Anil Kumar Singh

Abstract

Provides an insite into the basic concept of e-journals and discuss the advantages .This paper focus on e-journals concept, need of e-journals, types of e-journals, accessibility, merits and demerits of e-journals. The paper also explains the importance of a consortium of the various institutions, Universities. Problems faced by the libraries in handling e-journals are also discussed.

Keywords: *E-resources, E-document, E-journal, UGC-Infonet, INDEST Consortium.*

Introduction

An electronic journal is defined as "any journal, magazine, webzine, newsletter or type of electronic serial publication which is available over the Internet". Electronic journals can be access by using different technologies such as the world wide web (www), gopher, ftp, telnet, e-mail or listerv. Most of the all modern e-journals are available via the web (http://www.awe.ac.uk). With increasing the popularity of electronic resources, library and information centers migrating from traditional libraries with only print sources to digital libraries. The implementation and application of Information Communication Technology (ICT) in library and information centers provided nobel opportunity for development, storage, processing, accessing and dissemination of information. Electronic journals are most popular among scholarly research. Electronic resources have increasingly become the focus of Research and Development in the recent years. The web based full text electronic resources have become the most popular tools for library users. Internet facility to access the electronic resources. Most of the print resources such as journals are textbooks, reference books are expensive, there is rise in the subscription price of journals and database on an exponential rate.

Why E-Journals?

- Allow remote access.
- Can be used simultaneously by more than one user.
- Provide timely access.
- Support searching capabilities.
- Accommodate unique features such as links to related items.
- Save physical storage space.
- Contain multimedia information.
- Do not require physical processing, e.g., receiving and binding
- Can be environmentally valuable.
- Can be saved digitally.

Features of Electronic Journals

The following features of e-journals as comparision to print to print journals are as:

1. Remote Access: publishers offer the right through one of the following modes;
   a) User ID and password
   b) IP enabled (Intranet) Libraries who have Intranet based LAN.
   c) Combined: above both methods are also by publishers.

2. Number of users can access simultaneously at a time.
3. Timely access.
4. Supports various items, reference linking.
5. Save physical storage space.
6. Multimedia supports.
7. Do not required physical processing (receiving, binding etc.)

Types of Electronic Journals

There are three types of electronic journals are available.

(i) ***Online Journals***—Available on 'Pay-as-you go' or 'cost-per access basis' via online host or vendors. Online basis are OCLC, Dialog, Bibliographic Retrieval Services (BRS), STN International etc. Electronic version of the same existing printed journals are also available. e.g. American Chemical Society through STN International.

(ii) ***CD-ROM Journals***—Available in full text form published and distributed .Mostly they are electronic version of printed journals and published in CD-ROM. The most important journals and conference proceedings of IEEE, ABI -INFORM a full text business periodical published by University Microfilm International and ADONIS published by Elsevier publishers.

(iii) ***Networked E-Journals***—These are available only in electronic version over the network. Many printed journals are available in electronic form over Internet either free of cost. Many of the journals have no print version and are available only on the Internet.

The networked e-journals can be distributed by the following ways.

a) ***Content Page and Abstracts*** : The control computer holds a site of subscribers and send them content page and abstracts of articles by e-mail of the new published issues.

(*b*) ***Full Text Delivery*** : Subscribers are automatically sent the full text of each newly published issued by e-mail.

(*c*) ***Client/Server Technology***: A number of journals published with the application of client/server technology e.g. journals can be accessed via Gopher.

Availability of E-Resources

- E-Journals- full-text
- E-Journals-Abstract
- E-Books
- E-News papers
- Encyclopedia online
- Encyclopedia CD based
- E-Thesis and Dissertations
- E-Dictionary
- E-Newsletters
- E-Conferences proceedings
- E-Patents
- E-Reports
- E-Standards

Consortia Based Accessibility

An established consortia is a group of organization which regularly purchase electronic full-text databases as a unit usually located in one geographic area and often has a co-ordination who acts as a spokesperson for the group. Different models of consortia are as under-

(*i*) ***Print plus Electronic Access*** : The total holdings of a consortium is up-to an acceptance levels a negotiable payment will provide electronic access to the full-text database.

(*ii*) ***Electronic Only*** : Each Institution purchase the database for a fixed form and discounts are negotiated depending on the number of Institution joining.

(*iii*) ***Flip Pricing*** **:** A price is calculated on the predicted usage and value to each Institution. Print is available to purchase at a deep discount and is outside the agreement giving the library the option of cancelling titles when necessary.

Full-text Database

1. Institution wide license-members can access both On and Off –site
2. Alerting Service- Personal e-mail notification of a new articles keep users up-to-date with the latest research.
3. Reference Linking- Facility to move easily and swiftly between related articles.
   - American Chemical Society(31) http://www.pubs.acs.org
   - American Institute of Physics(18) http://www.aip.org
   - American Physical Society(11) http://ww.aps.org
   - Annual Reviews(29) http://arjournals.annualreview.org
   - Cambridge University Press(189) http://www.journals.cambridge.org
   - Elsevier Science(34) http://www.sciencedirect.com
   - Emerald(30) http://www.emeraldinsighc.com
   - Taylor and Francis(1105) http://www.journalsonline.tandfco.uk
   - J-Gate Gateway Portals(15000) http://www.j-gate.inforindia.co.in/
   - Encyclopedia Britannica http://www.search.eb.com

Consortia Based Service in India

- **CSIR** (Council of Scientific and Industrial Research) consortium http://www.niscair.res.in/activitiesandservices/majorproject/majproj.html

- **FORSA** (Forum for Resource Sharing in Astronomy and Astrophysics) consortium http://www.iiap.ernet.in/library/forsa.html
- **IGCAR** (Indira Gandhi Center for Atomic Research) consortium
- **IIM** (Indian Institute of Management)consortium
- **INDEST** (Indian National Digital Library in Science and Technology) consortium http://www.library.iitb.ac.in/indest
- **UGC-Infonet** consortium http://web.inflibnet.ac.in/info/ugcinfonet/ugcinfo.net.jsp
- **URLC** (Urdu Research Library Consortium) http://www.dsal.uchicago.edu/cgibn/urlc.py

International Consortia

- Illinois Digital Academic Library Unit, http://www.idal.illinois.edu
- **(ICOLC)** International Coalition of Library Consortium http://www.library.yale.edu/consortia
- Marine Academic Library Consortium. http://www.malc.marin.cc.ca.us
- New York Online Virtual Electronic Library. http://www.nysl.nysed.gov/library/novel
- North Caroline AHEC Digital Library and Resources System (ADLRS) http://www.nsl.unc.edu/ahec/adlrs/adlrshome.html
- South Asia Library Consortia. http://ww.libuchicago.edu/e/su/southasi/soab.html
- Library Consortia in Oregon, Washington and Idaho http://libweb.uoregon.edu.il.us/rbls/ql
- Washington Research Library Consortia http://www.wrlc

Merits and Demerits of E-Journals

- Increase the cost benefits per subscription and rational use of funds.
- Ensures the continuous subscription of the periodicals.
- Develops technical capabilities of staff in the using e-resources.
- Provide access to wider number of e-resources of lowest cost.
- Consortium with its collection purchase option has attracted highly discounted rates.
- Rates offered of e-journals through consortium are lower at 50% to 90%.
- Research output is expected to improve with increase access to international databases.
- E-subscription initiatives will bring remarkable changes in sharing resources.
- Provides archives access to the collective and also backup facility.
- Benefits librarians in accessing more literature save lot of space.
- Access to wide range of electronic database.
- Optimum utilization of rare collections.
- Easy access to resource sharing on Internet.
- Sharing of resources through union catalogues.
- Shared investments and time consuming.
- Reduce information cost.
- Increases cost- benefit per subscription.

Limitations of E-Journals

No Ownership : Print journals has its ownership but it is not possible in case of e- journals. The concept of "*library has*" or "*library has not*" is no more in existence for e-journals. Most of the publishers of e-journals maintains a limited periodical (say one year) for accessing them and accessing beyond that period paying extra charge.

Infrastructure : Subscription have to require necessary computer with facility to access, download and preserve.

Archiving : Today's a number of library are not updated with new technological applications. It requires large disk space to store and archives even if the data are stored, retrieving and providing printouts on demand, upgrading the retrieval software etc. are some of the problems that the libraries encounters.

Standardization : E-journals generally lack standardization regarding its format and the software associated in accessing them. Some of the e-journals do not include page numbers as the size of the page varies according to the display area of different computer (except pdf file).This raises citation problem when the same material to be cited can appear id different pages.

Acceptabilty : Many users are strongly in favour of paper of paper media.

Mode of Subscription : Some e-journals publishers offers free trial subscription for a period. Even for browsing an e-journal the user has to pay for it which the user may not always like to do.

Scrolling : The full page cannot display at a time and are need to scrolling of the screen which is very uncomfortable for reader.

Bibliographic Control : No suitable bibliographic control methods are available for e-journals thus it creates difficulties to identify a particular e-journals.

Suggestions

1. Do not opt for e-journals unless your library has required infrastructure, such as high bandwidth internet connectivity, LAN, storage device such as Snap server or juke box.

2. Do not discontinue subscription for Print journals in favour of e-journals unless e-journals are made available on CD-ROM/DVD with frequent updates.
3. Select only less frequently used journals in electronic form to begin with.
4. Prefer electronic form if journals are required for a time bound project.
5. If there are branch libraries ask for consortium subscription.
6. Insist on access for e-journals through IP address rather than password, if your organization has static IP address.
7. Give access to all the e-journals through the library homepage (both on intranet and internet).
8. Checkup the restrictions such as downloading, printing, re-distribution and royalty to the publishers in case of re-distribution.
9. For storing and retrieving in-house collection of e-articles prefer software available in public domain.

Conclusion

E-Journals are being added to library collections at exponential rates. Libraries do not extensive work to make electronic resources available with the need for individuals to enter the library. Recent developments in e-journals offer flexibility and enhance the visibility of e-journals by integrating all journals title in any format into a single list. E-journals are quickly becoming a mainstream form of scholarly communication. E-journals are facing many challenges but greatest benefit is the easy availability on the web which can provide immediate access. E-journals being relatively a new trend in the information world has generated lot of debate over its access, storage, preservation, and copyright. It is still evolving and only the time will tell us what shape it would take. Its further development and issues surrounding it can be resolved with collaborative efforts of librarians, researchers and the publishers.

References

1. **Mounissamy, P. and Swaroop Rani, B.S.** "*Evaluation of Usage and Usability of E-Journals*". SRELS Journals of Information Management. Vol. 42, No. 2, June 2005, Paper P. p189-205.

2. **Neelam Thapa and Sahoo, K.C. et. al.** "*E-Publishing: Beginning of a New Era of E-Journals*". IASLIC Bulletin Vol. 47, No. 4 December 2002.

3. **Krishnamurthy (M).** "*Consortia Based Resources Sharing and Accessing E-Journals*". Information Studies. Vol. 13, No. 3, July 2002.

4. **Dotty Lobo Janet and Bhandi, M.K.** "*E-Consortium in the Digital Era*". SRELS Journals of Information Management. Vol. 44, No. 1, March 2007 Paper F. p 63-76.

5. **Anil Kumar, N.** (2201). "*Electronic Journals : Major Issues*". Information Today & Tomorrow, 20(30),23.

6. **Kumber, T.S. and Karisiddappa, C.R.** (2004). "*Electronic Journals in Information Technology Application in Libraries: A Textbook for Beginners*". Edited by M. Mahapatra and D.B. Ramesh. Bhubaneswar: Reproprint. pp 293-97

7. **Martinez, Micheal L.** (2005). "*The E-Publishing Channels*". http://www.micheal-martinez.com

12

IGNOU—Reaching to Unreached with E-Learning

Sher Singh

Introduction

The Indira Gandhi National Open University in its 25 years of existence has emerged as the single largest University in the democratic world. It serves the educational aspirations of more than 3.0 million students in India and 36 other countries through the 21 Schools of Study and a network of 62 Regional Centres, more than 3000 Learner Support Centres and around 53 Overseas Centres. The University offers 338 Certificate, Diploma, Degree and Doctoral programmes comprising around 2000 courses. Conventional teaching -learning methods are being effectively supplemented with Information and Communication Technology and Satellite-based teaching-learning systems.

The University provides multi-channel, multiple media teaching-learning packages in the form of self-instructional print and audio/ video materials, radio and television broadcasts, face to-face counselling/tutoring, laboratory and hands-on experience, video conferencing, interactive radio counselling, interactive multimedia CD-ROM and internet-based learning. Apart from the print based self- instructional material, the educational programmes are reaching over 5 million homes through the Gyan Darshan Channels, via the DTH (Direct-To-Home) platform and web casting. The University is now gearing towards the development of interactive multimedia content and learner support through web-based platforms.

The e-Learning market is booming world over and is predicted to follow an upward swing with more and more institutions, organizations and individuals implementing and adapting to this mode. The power of e-Learning lies in its potential to provide the right information to the right people at the right time, place and pace. With advent of Information and Communication Technologies (ICTs), the delivery of educational programmes has witnessed a paradigm shift from print based teaching-learning to e-Learning. The learners having access to the internet and those who have adapted to learning through computers have shown their preference for e-Learning over print.

Web that gives information access to users who are physically remote from resources is emerging as a democratizing, emancipating, empowering force facilitating self-publishing, knowledge sharing and peer-to-peer networking. It has now shifted from being a medium, in which information was transmitted and consumed, into a platform, where content is created, shared, remixed, repurposed, and passed along. In the same spirit e-Learning has moved from being merely a content repository and emulating classroom teaching to more dynamic concepts of Social Networking, Do-It-Yourself (DIY), Personal Learning Environment (PLE) and Mobile Learning.

The ICT led initiatives in the form of e-Learning, online student support, digital repositories, open source courseware, etc. are now part and parcel of the Open and Distance Learning (ODL) systems. The earlier generations of ODL has given way to new generations which are ICT dependant for dissemination of knowledge without compromising on the quality and being more cost-effective.

e-Learning Initiatives at IGNOU

Realising the potentials of e-Learning to reach out the un-reached, IGNOU has recently embarked on a few major initiatives towards developing online learning environment for distance learners.

e-GyanKosh : It is an effort towards developing a national digital repository of learning resources. Over the years, a vast amount of self-instructional materials, audio and video programmes have been generated by the open and distance learning (ODL) institutions in the country. Huge wealth of structured knowledge is available which is very difficult to access and reuse. This results in duplication of efforts in developing learning materials and wastage of valuable resources. These learning materials have restricted use by the registered programme specific audience only. Some of the materials especially the audio visual ones do not even reach the targeted audience. Considerable amount of money is invested for the purpose but most of the materials remain under-utilized. Moreover, the audio-visual programmes stored in analogue form have a limited life span. Keeping these issues in mind, eGyanKosh, a national digital repository was initiated by IGNOU to store, index, preserve, distribute and share the digital learning resources developed by the ODL institutions in the country.

e-Learning has moved from being merely a content repository and emulating classroom teaching to more dynamic concepts of Social Networking, Do-It-Yourself (DIY), Personal Learning Environment (PLE) and Mobile Learning

The University is surging towards developing complete virtual learning environment to facilitate reaching the un-reached especially those located in far flung areas. From the management angle, a decision has been taken in the IGNOU's Planning Board to incorporate online delivery of its new academic programme parallel to the print based mode.

While certain learning functions can be best performed by Learning Management Systems (LMS) based on structured content, higher learning that is based more on discovery and exploration require learner-centric, pick-and-choose tools. E-Learning technology can be put to good use by enabling blended learning through information, interaction and collaboration. Therefore, apart from the course based structured e-Content, an e-Learning environment requires the following features: personal knowledge management tools- RSS feed, bookmark, tagging platform to Interact with the instructor and connect with fellow learners- e-Mail, chat or VoIP, discussion forums, blog, social networking (Orkut.) constructivist content development platform- Wiki platform for learner expression- blogs or portfolios.

Most of these tools are now available as open source, which facilitates easy integration with the existing e-Learning platforms. The platform for e-Learning delivery at IGNOU is now being developed integrating third generation e-Learning tools. An agreement has been signed with Google Inc. to integrate Google Apps with the system facilitating collaboration and real time interaction.

LIVE (Library and Information Virtual Education) : It is an initiative of IGNOU to develop an in-house Learning and Content Management System for imparting online education. It is envisaged as a complete virtual learning environment suite covering all the activities from registration to certification. In the first phase only the Master's Degree in Library and Information Science (MLIS) is being announced at international level. The virtual learning environment has the following features and processes involved in it.

Walk in Admission : Admission will be available throughout the year. The Registration form has been specially devised with the facility of uploading scanned certificates, other required documents and photograph. Registration will be confirmed only after verification of the certificates and payment of the required admission fee. Facility for online payment gateway is being integrated for the purpose.

Integrated Multimedia Courseware : once registered, learners will have access to personalised learning space (My Page).

This includes self instructional material, related audio/ video, slides, self check exercises, etc. weaved in one platform.

Online Counseling and Mentoring : Webcast based counselling integrated with text based chatting facility to be used for counselling purposes. Distributed model is to be followed with e-Counsellors available in remote locations.

24 × 7 *learner Support* : Asynchronous and synchronous modes of interaction will be used to provide just in time support to learners.

Portfolio Based Continuous Evaluation : Portfolios with individualised randomly generated assignments (both objective and subjective) automatically generated from question bank. Portfolio space for keeping track of course status, assignment status and grades.

Assignment Management System : In the first phase a question bank has been created with five hundred questions in each course. In the next phase web ontology/semantic web methods are to be used for automatic question generation. Complete automated system will be used for creation of individualised assignments and further distribution of completed assignments to concerned counsellors for evaluation. Assignments will be of two types, objective and subjective. Objective part will be evaluated automatically by the system and subjective part by the e-Counsellors online.

e-Tutor based practical- online instructions and e-Tutorials for hands on practice of WINISIS and Library Automation Packages. Multimedia manuals for self learning will be also provided.

Group Based Online Seminar : Blogs, chat rooms and discussion forums are to be used for group based seminars. Possibility of integrating EDUSAT based videoconferencing facility is also being explored for Seminar component.

Online Project Platform : Templates developed for synopsis submission, project uploading and evaluation. Viva voce will be conducted through Skype or through web based video conferencing tool.

Online Term End Examination : same pattern to be followed as in the case of assignments. However, examination will be conducted in specified centres for proper monitoring.

The same platform is now being replicated for running other programmes of the University. The MLIS programme will be followed by Post Graduate Certificate in Cyber Law in near future.

Sakshat : One Stop Education Portal, a landmark initiative of the Ministry of Human Resource Development (MHRD) to address all the education and learning related needs of students, scholars, teachers and lifelong learners has been developed at IGNOU. The portal envisages providing one stop solution to educational requirements of learners ranging from Kindergarten to higher education and education covering all fields of study including vocational education and learning for life skills. The pilot was inaugurated by the President of India Dr. APJ Abdul Kalam on October 30, 2006. The portal has three tier architecture with user interface, content management system including built in learning object repository and administrative module. The user interface provides basic five modules viz. educational resources, scholarship, testing, super achiever, and interact. Apart from this, it has an inbuilt virtual class which follows four quadrant approaches, facilitating single window topic based access to learning material. The 'interact module' provides facilities like talk to a teacher, blogging, discussion forum, career counselling, etc. The content generation module follows a Wiki approach of content development and deployment facilitating storing of content based on metadata in the repository.

PAN- African E-Network : PAN-Africa e-Network Project of the Government of India IGNOU has been given the responsibility of meeting the educational requirements of African learners in all the countries of the African continent through the tele-education mode. Under this pilot project, IGNOU has signed an MOU with the Universities of Addis Ababa and Haramaya,

Ethiopia for offering the IGNOU MBA programme to the students of these Universities. The forty students admitted in the project are taught by IGNOU's core faculty through e-network system, established by TCIL. Efforts are on to extend the project for other programmes of the University and develop SCORM compliant content for e-Learning delivery mode.

As the University is surging ahead with e-Learning mode of delivery the real challenge is to develop e-Content at mass scale. As more and more programmes will be offered online in future, immediate attention needs to be paid for preparing the faculty for digital environment both in terms of content generation and programme delivery. The sustainability of the online programmes will depend on the accessibility to the basic infrastructure on the learner side. University has initiated a major effort in this front by equipping the regional centres and study centres with necessary hardware/software and bandwidth.

Sustainability of the online programmes will mainly depend on the equity and access to rich learning content and 24 × 7 online support. Mere duplication of print material into e-Learning content will in no way be beneficial to the learners.

Conclusion

An integrated approach is needed where in one window operation learner has access to course content, and other learning resources in the form of e-Books, e-Journals, educational multimedia programmes, radio and TV channels, etc. along with complete support service. The University recognises this and is trying to harness the capabilities of the ICT in education to fullest extent to reach out the un-reached.

References

1. **Asian JDE** (2003). "*Asian Journal of Distance Education*", Tokyo: Fukuka.
2. **Nayar, D. P.** (2009). "*Education for Rural Development*", New Delhi: Shipra Publications.

3. **Gupta, A.** (2008). "*Education in the 21st Century*", New Delhi: Shipra Publications.

4. **Ramanujam P.R.** (2006). "*Globalisation, Education and Open Distance Learning*", New Delhi: Shipra Publications.

5. **Vetukuri P.S. Raju** (2007). "*Education of the Masses*", New Delhi: Shipra Publications.

6. **Ramanujam P.R.** (2009). "*Distance Open Learning*", New Delhi: Shipra Publications.

7. **Hazra, A.** (2010). "*Rural Literacy in India: The Changing Scenario*", Kurushetra, 58 (11), 3-4.

8. **Negi U.R.** (2003). "*Student Support Services in Correspondence Course Institutes in Universities*", New Delhi: Association of Indian Universities.

9. Careers 360 (2010) "*Doorasth Sikshan Sansthan*", New Delhi: Outlook Publishing.

10. **Kanjilal, Uma** (2008). "*Digital Learning*", New Delhi.

11. Websites-www.ignou.ac.in

13

E-Resources for Art and Culture in Indian Scenario

Arupa Majumdar

Abstract

Libraries function as an essential integral component in higher education system. Advent of information and communication technologies and their capabilities such as high-resolution capture devices, dramatic increase of digital storage media, explosive growth of Internet and WWW, sophisticated search engines, fast-processing power and reducing cost of computers, high bandwidth networks and increasing number of electronic publications have made possible to switchover to technological solutions to the problems of present day university libraries which cannot be won otherwise by the traditional system of librarianship. The present paper focuses of electronic resources for art & culture illustrates the advantages of e-recourses in the light of the major problems, issues.

Introduction

Library is a repository of resources that create an intellectual environment in a education system and with the aid of internet technology it changes the whole scenario of information system from print to electronic which empower and enrich the higher education system. Libraries began creating portals as a virtual window as well as digital libraries steaming from their print collections. Thousand of public works literary and scientific articles, pictures and sound track became available on the screen and users are able to access e recourses either by local or remote locations. The proliferation of electronic resources had a significant impact on the way of academic community uses. Libraries prefer electronic resources to substitute collections for optimum use. In the present digital era dissemination of e-resources to the scholastic society and community has become an important task though information is transmitted through various media such as periodicals, serials, journals database to academic community. However in Indian scenario it observes that the scope of e-resources are more inclined and popularized towards the field of scientific information like e resources in science & technology in India are mainly promoted by CSIR, HELNET, FORSA, IGCAR, JCCC & VIC, INDSAT, SONET, OUHYD Math consortium etc. these are popular among the academics, as compare to that e resources of art and culture are less know to the academicians although there are few setps are taken towards the development of E-Resources for Art and Culture. The present paper is discusses about the importance of electronic recourses for art & culture which open up many exciting features for academic libraries with both opportunities and challenges.

Defination

Electronic Resources may be defined broadly as any journal, magazine, e-zine, webzine, newsletter or type of electronic serial publication which is available over the Internet and can be accessed using different technologies such as World Wide Web (WWW), Gopher, ftp, telnet, e-mail or listserv. The University of Glasgow defines the term 'Electronic recourses' as "Any peace of information that is available over the Internet can be called an 'electronic resources'. In some cases, print equivalents exist; in some cases, not. Some electronic resources are freely available; other has charging mechanisms of different types. Established publishers issue some; others are produced from an individual academic's office. In simple words it is said that the electronic resources are the one where the text is read on, and /or printed from, the end-user's compute rather than as print-on-paper. In the online information, the data is downloaded directly from the host computer. Define it strictly to be "a full text electronic publication, which may include images, and is intended to be published indefinitely". So it is said that e-resources are publications available in digital format. Some are distributed on CD-ROMs, some over the Internet. Thus an Electronic resource is defined as the grouping of information that is sent out in electronic form with some regularity. It covers any serial or serial-like publication available in electronic format, which is produced, published, and distributed electronically. Networked electronic journals are based on mailing list software or client/server computing applications, including Gopher and WWW.

Art and Culture of India : Scope as a Subject

The vast scope of Indian art is deeply intertwined with the country's cultural history, religions and philosophies. Indian art, therefore, provide an aesthetic continuum that extends from the early civilization to the present day. From being essentially societal in purpose in the beginning, Indian art has evolved over the years to become a fusion of various cultures and traditions. Fuelled by a robust Indian economy, the energetic participation of young expat Indians, an increase in the number of collectors worldwide and a growing international attraction for owning these works, Indian art is enjoying a remarkable period in its history. The growth in this dynamic market has been characterized by increasing interest in more traditional Indian art forms as well as the continual growth of modern and contemporary art.

Significance of e resources of art and culture:

- It serve as integrated database which preserve the valuable information of Indian art & culture.
- Help researchers and archaeologists in searching and viewing the information of their interest.

- Bridging Digital divide
- Promotion of Cultural tourism

Indian Initiative in E-Resources for Art and Culture

In India, few Institutions felt for the importance and necessity of digital repository for art and culture and developing an online access of information, to foster the research needs and craving for information of the scholarly community. The organization like INGCA , Archaeology survey of INDIA with the aid of Ministry of communication & Information technology taking major efforts of electronic documentation of Indian cultural resources using digital technology.

National Databank on Indian Art and Culture

The project named National Databank on Indian Art and Culture: sponsored by the Department of Information Technology, Ministry of Communication and Information Technology (MCIT), and Archaeological Survey of INDIA, Government of India The main objective of the project is to enhance the accessibility of Indian cultural resources using digital technology. The project includes the digitization of information related to various aspects of Indian art and culture accessible from a single window. Contents including over 1 lakh visual, 1000 hours of audio and video, 25000 rare books on art and culture and walk-through of some of the archaeological monuments will be covered. This will be one of the major source of information on Indian art and culture, which can be accessed by the researchers, students, art historians, archaeologists etc.

We can access it on http://www.ignca.nic.in/ndb_0001.htm

Scope of the Projects :

- Digitization of images, audio, video uploads on the website for the academic reference to the researchers with their catalogue. This includes the data from the archaeological sites/monuments (covered under the ***ASI and state archaeology departments*** and large no. of monuments not covered under these departments) and from the **life style of communities** from allover India depicting various facets of arts.

- ***Digitization of Rare books*** : It is proposed to digitize 25000 rare books (estimated to be 50 lakh pages) @ 200 pages per book) on Indian Art and Culture from the ASI collection, which will be uploaded on ASI's website for public access.
- ***In addition to that*** it is proposed to create walk-through of selected archaeological monuments and sites by taking the stills and stitching the images through the software. For example *Brhadisvara Temple, Thanjavur, Tamilnadu, Keshava Temple, Somnathpura, Karnataka, Rani-ki-Vav, Patan, Gujarat, Tomb of Sultan Garhi, Nalikpur Kohi, Delhi, and Jain Temple in Old Delhi* etc. will be considered.

Kalasampada

Resource for Indian Cultural Heritage **(DL-RICH)** project sponsored by the Department of Information Technology, Ministry of Communication and Information Technology (MCIT), Government of India. Project aims to use multimedia computer technology to develop a software package that integrates variety of cultural information and helps the users (students, scholars, artists and Research & scientific Community etc.) to interact and explore the subject available in image, audio, text, graphics, animation and video on a computer in a non-linear mode, by a click of mouse. The system aims at being a digital repository of content and information with a user-friendly interface. The knowledge base such created will help the scholars to explore and visualize the information stored in multiple layers. This will provide a new dimension in the study of the Indian art and Culture, in an integrated way, while giving due importance to each medium. The IGNCA's archival collection includes manuscripts, slides, rare books, photographs, audio and video along with highly researched publications of books, journals and newsletters.

These rare manuscripts, collected from different institutions of India and abroad are mainly in the form of microfilms and microfiche. Rare Photographs digitized includes Rajah Deen Dayal collection, Sambhu Nath Saha collection etc. About thousand hours of Audio/Video Film & Video Documentation (online audio/video) is available in digital form out of total collection of over 10000 hours. A part of the IGNCA's publications (electronic books available) with the various volumes of News Letters (Vihangama) and Kalakalpa: Bi-annual Journal have been digitized. The major activities of the project includes the digitization of materials, post digitization editing, high capacity storage & backup system, designing and development of effective retrieval system etc. Technology used for this development is based on Open standards using Unicode, a multilingual standards for fonts, accepted worldwide with open type fonts. Search is available both in English and Hindi (Devanagari). User have the option to select the material of his interest either from a specific type of collection like books, manuscripts, slides, audio, video etc or from the entire collections. The facility is currently available only on intranet, for the very fact that these materials are priced possession, Intellectual Property Rights, and copyright issues, etc.. Although, partial information can be accessed from the IGNCA's. Available Metadata of the contents will be finally converted in Extended Dublin Core, as practiced for majority of the digital library projects worldwide. Digitization standards followed at the IGNCA for the digitization of various kinds of materials are given below. These also conforms the UNESCO Guidelines published in April 2002.

We can access it on : http://www.ignca.nic.in/dlrich.html

CoIL-Net–Content development in Indian Language Network

It is a project sponsored by Ministry of Communication and Information Technology (MCIT).

- To enhance the access to cultural resources using digital technology.
- To develop a reusable 'MODEL DESIGN' and 'Development Process' for implementing user friendly web

enabled heritage library for HINDI SPEAKING POPULATION AND OTHER HINDI KNOWING PERSONS IN INDIA & abroad.

- To implement a web enabled Hindi Based multimedia heritage library also offering contextual and vetted links to important websites to contribute towards the socio-economic development of Hindi Speaking region.
- The digital library shall offer carefully selected, thoughtfully compiled and contextually integrated multimedia content on cultural heritage, folk literature and life style of Hindi Speaking region specially the states of Uttar Pradesh, Madhya Pradesh, Bihar, Rajasthan, Chattisgarh, Uttarakhand, Jharkhand.

Challenges

1. ***Financial Constraints*** : The infrastructure required to display, store or print electronic resources are expensive. Downloading and printing each article will be a costly affair. This means a net increase in economic and ecological costs and it becomes a relatively expensive way to acquire a single copy. Many e- resources do charge subscription fees. The pricing schemes of some suppliers are very complicated and limiting, and this might hinder libraries from utilizing e- resources.
2. ***Stability and Storage*** : The volatility of e- resources makes preservation of e- resources a major concern. In case of the benefits of access are enhanced, the ability of electronic resources to transmit information through time is not completely confirmed. Offline storage methods suggested are magnetic media, such as tape, hard disks, and floppy disks, and optical media such as CD-ROMs. There are issues of preservation of storage media, hardware and software dependency and dynamic versions of electronic resources that also need to be dealt with.
3. ***IPR Issues*** : There are certain issues like; protection of the intellectual property of the author in order to preserve

the originality and integrity of the work; warrant for the attachment of the author and the work in public; protection of the author's ideal and economic interest and benefits, including publication and reproduction of his/her work. This is usually accomplished through the publishers, who disseminate the work in an appropriate, protected and retraceable manner. Electronic resources presently emphasize information access instead of ownership

4. ***Social Constraints*** : Electronic interfaces can take a long time to master. Electronic searching, downloading and printing replace the traditional activities of physically browsing, scanning and photocopying resources articles. The intricate steps to accomplish the previously simple or habitual tasks might frustrate users. People read up to 25 to 30 percent more slowly on a computer screen than on paper.
5. ***Technological Constraints*** : Digital resources depend up on technology and equipment for storage and display. Proper Infrastructure facilities are required for the access. The academic community can be divided into "haves" and "have-nots" because of access to equipment and network. The network or connection speed can be too slow. Screen quality of graphics and photos is still primitive when compared to print media.
6. ***Users' Access*** : Depending on the licensing agreement and local funding, downloading and printing can be provided in libraries as well as at the desktops of the users. Minimum hardware and software requirements are going to progress as technology progresses, but basic entities such as hard drives, colour monitors, external disk drives, printers, security cables, tables and chairs are often inevitable to be equipped onsite. Internet connection and bibliographic linking software are extras to provide value added service.
7. ***Training and Support for Staff and Users*** : With the number of e- resources being published and the variety of

different interfaces, more sophisticated searching and retrieving skills are necessary. If library staff is provided with adequate training and support in order to be aware of new development of technology, more flexible and suitable services can then be available for users. The information provider role of libraries remains important but the delivery and type of services might have to adapt to the changing technology and users' needs.

Conclusion

It is said that E-resources can't replace print resources yet because only a fraction of scholarly materials is available electronically. What is available varies in quality, accessibility and price. But E resources provide many opportunities and potentials for academic libraries. Out of the advantages and disadvantages of e- resources librarians need to be able to identify and balance the factors that would make e resources a success or failure in their libraries. Certain bottlenecks still arise with e resources such as users experience frustration and difficulty in their first efforts to use e-resources if they lack proper infrastructure they may oppose efforts by libraries to replace printed resources by electronic ones. Librarians facing financial pressure identify journal price rise as a significant contributing factor. Library users want the advantages of the digital format, but until archiving issues have been satisfactorily addressed, many librarians consider it necessary to acquire the print format as well. As a result, total subscription fees and delivery costs have increased significantly. Developing a common vision of the future of IT can consolidate efforts to tap into the evolving telecommunication infrastructure. There is a general consensus that e-resources would not replace but coexist with the print format.

References

1. **Woodward, Hazel etal.** "*Electronic Journals : Myths and Realities*". OCLC Systems & Services, vol.13, No.4, 1997. pp. 144-151.

2. **Chan, Liza.** "*Electronic Journals and Academic Libraries*", Library Hi-tech, vol. 17, No.1, 1999. pp. 10-16.
3. **Halliday, Leah.** "*Progress in Documentation Developments in Digital Journals*". Journal of Documentation, vol. 57,No. 2, 2001. pp. 260-283.
4. **Arora, Jagdish and Agarwal, Pawan.** "*Indian Digital Library in Engineering Science and Technology* (INDEST) Consortium : Consortia Based Subscription to Electronic Resources for Technical Education System in India : A Government of India Initiative. Paper Presented during CALIBER-2003, during 14-15, February 2003. pp. 271-290.
5. **Mange Ram et. al.** "*Digital Preservation : A Challenge to Libraries*". Library Progress(International), vol. 25, No.1, 2005. pp. 43-48.
6. INFLIBNET Centre, Ahmedabad, http://www.inflibnet.ac.in
7. INDEST Consortium Website. http://paniit.iitd.ac.in/indest/
8. Health Sciences Library & Information Network (HELINET) http://www.rguhs.ac.in/hn/newhell.htm
9. **Government of India.** "*Ministry of Human Resource and Development*". Committee of Experts on
10. Consortia-based Subscription to Electronic Resources for Technical Education System in India. New
11. Delhi: MHRD, 2002. P. 9 - 10. (Unpublished Report).
12. E-subscription under UGC-Infonet-Consortium-Update. INFLIBNET
13. **Quoted in Lee, Stuart D.** "*Building and Electronic Resource Collection : A Practical Guide*". London: Library Association Publishing, 2000. P49. Also available at :http://www.lib.gla.ac.uk/enquiries/farwhat.html#E-journal>, [Accessed on 10.12.2003].
14. http://www.ignca.nic.in/ndb_0001.htm accessed on 23/3/11
15. http://www.ignca.nic.in/dlrich.html

14

Role of Academic Libraries in Access to E-Resources

Manoj Kumar Dwivedi &
Rakesh Kumar Mishra

Abstract

The importance of Academic libraries depends upon the quality of the size of documents. The objective of academic library is to develop its documents qualitatively adequate and quantitatively relevant, i.e. the needs of users of library. Library & Information Centers facilitate the process storing, processing and disseminating of information Sources with the help of Information & communication technologies (ICT). The library professionals are utilising ICT to keep pace with the problem of informal instant access to digital information is the most distinguishing attribute of this paper, the authors tried to highlight the library services of different components of providing digital information services, access to sources and IT's role in modernising libraries.

Keywords: *ICT, Libraries, Electronic resources, Non-print Media*

Introduction

Before the advent of information and communication technology (ICT), academic libraries were the sole custodians of information, which was predominantly in print. ICT brought changes necessitated by new information packaging. Academic libraries are faced with managing hybrid resources (print and electronic) and are challenged to acquire the necessary skills. Furthermore, electronic information is eroding the monopoly of academic libraries as the sole access point to information. Nevertheless, academic libraries can maintain their place by serving as an access point to both print and electronic resources. This paper discusses the nature of academic libraries in the digital age including resources, the concept of universal access, and the role of the in universal access to print and electronic resources. It also presents and describes a conceptual model of resource access for academic libraries in developing countries.

Academic Libraries in the Digital Age

A well established academic library is essential for any academic institution. As a focal point for teaching, learning, and research, it is expected to provide standard information resources. Today, academic libraries are struggling to keep their place as the major source of inquiry in the face of emerging digital technology. Digital technology has revolutionized not only the way information is packaged, processed, stored, and disseminated, but also how users seek and access information. Academic libraries no longer restrict themselves to print services such as collection development, cataloguing and classification, circulation and reference services, current awareness, selective dissemination, and other bibliographic services, but have extended their efforts to interdisciplinary concepts and computer software and hardware and telecommunication engineering and technology. As observed by Campbell (2006:17), "numerous creative and useful services have evolved within academic libraries in the digital age: providing quality learning spaces, creating metadata, offering virtual reference services, teaching information literacy, choosing resources and managing resource licenses, collecting and digitizing archival materials, and maintaining digital repositories". Academic libraries presently are faced with not only the decision on what books and journals to acquire to satisfy faculty and students but also on how to remain relevant in the digital era, mindful of low budgets and resentment on the part of institutional administrators. There is also the issue of library users opting for alternate, more convenient, and "qualitative" sources of information i.e.Internet. As Lombardi (2000) notes, users will prefer more computer content, more and more computer indices, digitized finding aids, digital repositories of articles, online access to newspapers, etc. Libraries also struggle with when, how, who, and where to begin digitization efforts, while keeping in mind that hesitation in the digitization of institutional archives will result in relinquishing the function to another institutional repository host. The consequence is repositioning of academic libraries resources, operations, services and skills. Resources today occur in hybridized form: print and electronic, and therefore services provided and skills possessed by professionals in these libraries should reflect that trend. In the ICT age following factors are important for the advent of non-print media, which are as-

(*i*) Radical change in the methods of dissemination of information.

(*ii*) Advantages in knowledge absorbing had higher power of retention.

(*iii*) Embodiment of knowledge-retention, handling convenience.

(*iv*) Technological impact

(*v*) Socio-economic implications.

Universal Access to Resources

Academic Libraries have always served as access points for information. Services have evolved from the days of closed stacks, through shelf browsing and card catalogues, punch cards, and OPACS, to the concept of open access and institutional repositories. This historic migration has tried to satisfying the changing needs of library users, including ease of access, interaction richness, low interaction, and low cost. Access is more important than ownership

by Eisenberg (1990). The underlying issue becomes the provision of information resources in offices, hostels, classrooms, homes, etc., regardless of where the information is found.

Recognizing the importance of a new mode of information access, academic libraries took responsibility for automation. Funding bodies such as the Information & Library Network Centre (INFLIBNET), Autonomous body of UGC, India which pulls together resources electronically, connecting all the academic university libraries in India, with the hub at the INFLIBNET. The participating libraries become access points to the universal information resources.

Crow (2002) describes as institutional repositories as, "digital collections capturing and preserving the intellectual output of a single or multi-university community". An institutional repository is a way of reducing the cost of scholarly publication and increasing visibility and access of scholarly research from faculty and students of academic institutions by hosting them in the institution's, professional societies, or third-party provider's website. The institutional repository is a sort of mirror image of print institutional archives, and in some academic institutions is being maintained by the institution's library. While academic libraries were at the center of providing access to print archives, the institutional repository has given them the responsibility of providing access and also interoperability functions (standardizing metadata formats and metadata harvesting).

Conceptual Model of Print/Electronic Resources Access for Academic Libraries

The foregoing depicts a challenging situation for academic librarians who are expected to create universal access to both traditional and electronic resources. Hence, a model was developed as shown in Figure 1 to assist these libraries. The model consists of two access environments: the in-house, local, or independent environment, and the universal, global, or integrated access environment.

Three Alternatives for the In-house Environment

Figure 1 illustrates three alternatives. The first (A) is an e-resources Unit with the enabling environment (software, hardware, and trained staff), which is open to academic library users (2); A combined Print and E-resources Unit (B) with enabling environment (3)also open to academic library users (4); Print resources Unit (C) with enabling environment (5), open to academic library users (6).The three alternatives should be practiced in Indian academic libraries today. In as much as they are amenable to the

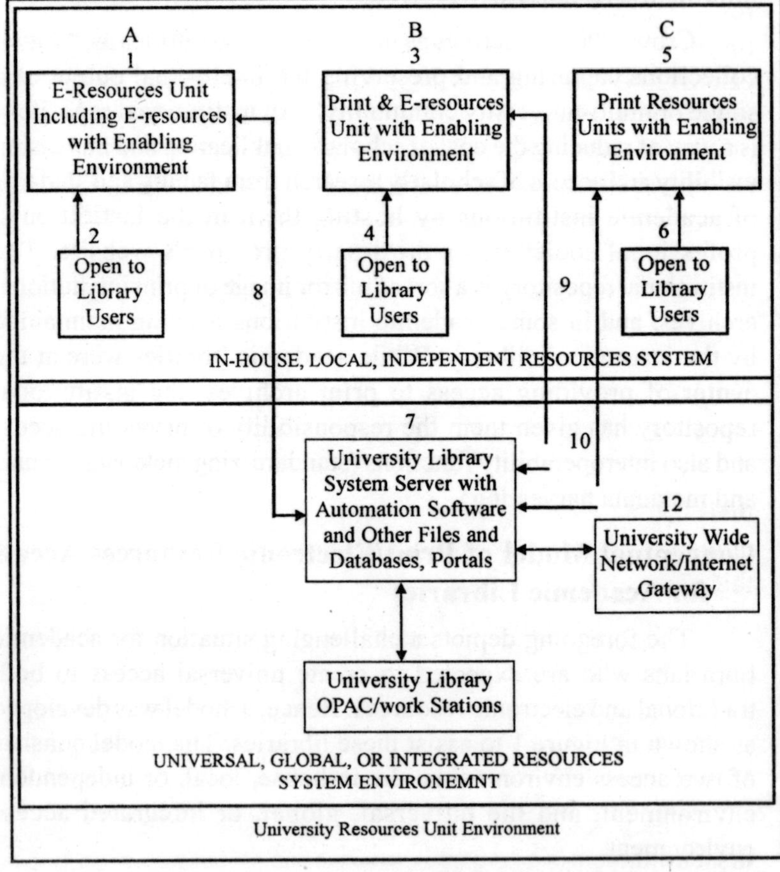

Fig. 1. Print/E-Resources Access Model for Academic Library

in-house access expected of traditional library services, they fall short of the universal access that is required of academic libraries of international standing. This environment is the hub of an electronic consortium. It works in a feedback mechanism with the e-resources, print and e-resources, or print resources, unit all of which revolve around the library server (7). It provides resources for the independent access environment and also receives resources from it.

The Server (7) with high capacity hardware, software, and telecommunication components, gets feedback from the e-resources unit (8), print and e-resources unit (9), or the print resources units (10). Each of the units can access the server, which can also be accessed universally through the library OPAC (11), available on the Internet (12). The integrated access environment allows academic libraries operating any of the options in the independent environment to be part of universal access with or without the involvement of their parent institutions.

The model consists of two access environments: the in-house, local, or independent environment, and the universal, global, or integrated access environment.

Emerging Issues for Academic Libraries in Developing Countries

For academic libraries to maintain a prominent position in their institutions, they must move from limited or local access to universal access. For academic libraries in developing to achieve this, requires expandability, flexibility and compatibility (Tebbetts 1991). It requires standard hardware, sufficient capacity, networking capabilities, flexible software, standards such as MARC for information storage and retrieval, local expertise, and a plan for the next system.

Conclusion

The main function of any library is to collect, store and disseminate information pointed by and exhaustively to the clientele. The collection, infant are the essences of library because it is what most users of library are after. The librarians are mainly there to build up the collections and help the users to exploit the resources to their most advantage. The emergence of information and communication technology(ICT) has repositioned the frontiers of academic library resources, operations, and services as well as expectations of user groups. The practice of walking to the library to consult the card catalogue and browse the shelves is moribund in developed countries, and this trend is quickly approaching developing countries as well. Academic libraries must embrace this scenario. The print/e-resources access model can serve as a stepping stone. When such a step is taken, academic libraries must remember expandability, flexibility, and compatibility.

References

1. **Campbell, J.D.** (2006). "*Changing a Cultural Icon: The Academic Library as a Virtual Destination*". Educause Review 41(1), 16-31.
2. **Cisse, C.** (2004). "*Access to Electronic Information and Information Research*". SCAULWA Newsletter 5(1), 14-17.
3. **Eisenberg, M.B.** (1990). "*Trends and Issues in Library and Information Science*". Syracuse, NY: ERIC Clearinghouse on Information Resources.
4. **Kim, B.B., & Whinstom, A.B.** (2002). "*Virtual Field Experiments for Digital Economy : A New Research Methodology for Exploiting an Information Economy*". Decision Support Systems 32(3), 215-231.
5. **Lombardi, J.V.** (2000). "*Academic Libraries in a Digital Age*". D-Lib Magazine 6(10) Available:www.dlib.org/
6. **Tebbetts, D.R.** (1991). "*Expandability, Flexibility, Compatibility: Key Management Considerations in Academic Library Automation*". Available:web.simmons.edu/~chen/nit/NIT'91/223-teb.htm

15

Future of Traditional Library Services in the New Online Environment

Navin Upadhyay

Abstract

Mobility, search and social networks have changed how the academic library patrons access information. Today's patrons except information in the palm of the hand. Rather than try to change users' habits, the library can change its approach and meet users where they are — on the Web, using the tools they enjoy using.

Keywords : Library services; Library 2.0;

Introduction

For some decades now the development of ICTs has been seen by most observers to be the most fundamental change of the period. Indeed, the question is being asked as to whether information and library services, in any recognisable form, will be needed in the new millennium. Would our visitor from a hundred years hence have any familiarity with the concept? More than fifteen years ago **Lancaster** predicted the demise of the library :

"Ultimately.... libraries as we know them seem likely to disappear. Facilities will still exist to preserve the print-on-paper record of the past, of course, but they will be more like archives, or even museums, providing little in the way of public service. As for the electronic sources, libraries may have an interim role to pay..... In the longer term, it seems certain that the libraries will be bypassed. That is, people will have very little reason to visit libraries in order to gain access to information sources".

Today, it is increasingly difficult to maintain and improve the usage rate of libraries due to the revolutionary changes in the technological, information and learning environment. Access to the Internet and availability of online resources is changing students, research scholar's and faculty member's information-seeking behaviors, the way they conduct research, and their perception and use of library services. These changes are particularly challenging for academic libraries that have to develop their collections and adopt their services in order to fulfill their mission to support teaching, learning, and research. Digital collections need not signal the death of the library, but to survive we need to develop a program that matches the needs of today's library user. Library services must address the needs of an increasingly online user. Services should focus on managing information resources and advanced applications of information technology, not on simple access.

Traditional Library Services

Traditional library service and the traditional library environment is highly structured. When one enters a library one enters a very orderly environment which operates according to a very sharply defined set of rules. If one does not understand these rules one has difficulty accessing any information. Traditional service is mediated... The patron does not, as a rule, experience or "find" the information for themselves. One goes to a librarian who performs the role of either finding or teaching the patron how to find the information or one uses a mediating source such a Print catalogue card cabinet/online catalog or database about which one must learn a series of "rules". What is important to note is that the library patron does not encounter or experience the raw information directly but only through a secondary source. Librarian is the central player in the traditional library and all activity focuses upon the librarian... The librarian interprets the Library for the patron Without the necessary prerequisite skills the patron will get nowhere in the library.

Patron is a passive receiver of information... In the traditional library environment the user or patron receives information from the librarian. Unless trained in the "rules" of the Library the patron is essentially a supplicant seeking the help of the authoritative librarian. In essence, the traditional model of library service as outlined above is a power relationship based on the dependency of the uniformed and incapable patron on the highly skilled librarian who knows the rules and is, thus, able to provide information to the unanimated.

Changing Information Seeking Behavior of User's

Over a decade ago, researchers, clinicians and information providers worldwide were enthusiastically embracing online access to information as a revolution in STM publishing. Indeed, web access to STM literature was hailed as a revolution on the order of Gutenberg's invention of the printing press. Today we are in the midst of a fascinating new phase of what clearly is the ongoing evolution of dissemination of scholarly content. We need just one word to sum up this new phase : mobile. Today's patrons except information in the palm of the hand. We are living in a transformational era of personal portable libraries in the hands of people rather than organizations. Mobility, search and social networks have changed how the academic library patrons access information.

Past practices of following traditional organizational guidelines are no longer acceptable to the patron. The practice of asking students and faculty to follow too many rules is being challenged, and demands to access information by using devices of choice are becoming commonplace. One would probably remember not too long ago the drive to make the Library the Third Place — after the home and the workplace. We must have been too successful because crowds in libraries are a common sight. But now the scenario has completely changed. Unfortunately the reality is that more than 90% of any survey would indicate that whenever the public wants to find out about something, the first thing they would do would be to "Google" it. No one would think about asking a reference librarian or even logging onto a library website to use the inquiry service.

This is the new marketplace; this is where most people live and work. Thus as libraries, we must recognize that the Internet and search engines are now the main ways in which people look for information.

Library 2.0 : The New E-world Order

The main characteristic of this new order is that information management and provision are no longer under the purview of the librarian. For example, cataloging may be a fundamental skill of librarianship but the art of social tagging on the Net turns that upside down because it is the reader who is now categorizing and defining information based on her or his own terms. And the best thing about social tagging is that everyone is allowed to categorize the information the way they want to. The phenomenon of Wikipedia breaks yet another golden rule of librarianship: that of content validation. We tend to believe that information is useful only if it is authenticated. However on the Net, everyone is in a position to provide and to collaborate on information. Not only are these platforms very popular, they are where people go to look for information!

Today, many libraries are at a crossroads: several of the services they have traditionally provided within their walls are increasingly made available online, often by non-library, commercial entities. For example, Web search engines provide easy access to millions of Websites, online databases provide convenient gateways to news and scholarship, and book-scanning projects make roaming the shelves seem antiquated. Meanwhile, the traditional authority and expertise enjoyed by librarians has been challenged by the emergence of peer-produced and collaborative knowledge projects, such as Wikipedia, Yahoo! Answers, or Amazon's personalized book recommendation system. Further, the professional, education, and social spheres of our lives are increasingly merging, marked by the rise of social networking services providing new interfaces for interacting with friends, the sharing of information, and professional collaboration.

Libraries face a key question in this new environment: what is the role of the library in providing access to knowledge in today's digitally networked world? One answer has been to actively incorporate features of the online – "Web 2.0" – world into library services, thereby creating "Library 2.0."

Although concepts from Business 2.0 and Web 2.0 had inspired Michael Casey to be the first library professional to coin the term Library 2.0 in 2005, other library innovators such as Michael Stephens, Stephen Abram, Phil Bradley, Meredith Farkas, and Laura Cohen have taken on the term and concepts from Library 2.0 and have broadened them to encompass a set of best practices.

Concepts of Library 2.0

1. ***Web 2.0 Technologies*** – increase the flow of information from the user back to the library, especially with online services like the use of OPAC systems. Browser + Web 2.0 Applications + Connectivity = Full-featured OPAC.
2. ***Remixing Library Services*** – Reaching beyond traditional boundaries of library space to push for services out to people in places where they are already interacting.
3. ***Constant Change*** – Adopting a strategy for constant change while promoting a participatory role for library users, harnessing the library user in both design and implementation of service.
4. ***Blurring the Service Lines*** – Harvest and integrate ideas and products from peripheral fields into library service models.

5. ***Experimentation***–Continue to openly and flexibly examine and improve services and be willing to replace them at any time with newer and better services
6. ***Discoverable Information***–The open library should seek to enable discovery, locating, requesting, delivery and use of its resources in its care
7. ***Culture of Participation***–Draws upon perspectives, suggestions, ideas, and contributions of all related to the library–staff, technology partners, patrons, and the public community
8. ***Collaboration***–For Library 2.0 to work, libraries, technology providers, policy makers, and information consumers must work together.

Library 2.0 is a change in interaction between users and libraries in a new culture of participation catalyzed by social web technologies. Interactivity can be interaction between the librarians and the customers or library users, but also between library staff or between users of the library services. This interaction enables participation where the users can contribute to the content of libraries and library services using new ways and new web-based tools or so called Web 2.0 tools. What makes a service Library 2.0? Any service, physical or virtual, that successfully reaches users, is evaluated frequently, and makes use of customer input is a Library 2.0 service.

Even older, traditional services can be Library 2.0 if criteria are met. Similarly, being new is not enough to make a service Library 2.0. Library 2.0 is all about library users—keeping those we have while actively seeking those who do not currently use our services. It's about embracing those ideas and technologies that can assist libraries in delivering services to these groups, and it's about participation—involving users in service creation and evaluation. Library 2.0 is an operating model that allows libraries to respond rapidly to market needs. This does not mean that we abandon our current users or our mission. It is a philosophy of rapid change, flexible organizational structures, new Web 2.0 tools and user participation that will put the library in a much stronger position, ready to efficiently and effectively meet the needs of a larger user population.

Conclusion

Mobility, search and social networks have changed how the university faculty and research scholar's accesses information. If you feel comfortable in a controlled environment and like in-person, one-on-one visits with patrons, your era of comfort is coming to an end. Today's society favors a quick online answer over a visit to the local library. The result is today's academic library needs to change to ensure it remains a relevant resource. The library is at a crossroads of choices to make in responding to the same mobile society. So libraries must give up control and make use of collaborative tools and technology to engage and share information rather than just provide it. Rather than try to change users' habits, the library can change its approach and meet users where they are — on the Web, using the tools they enjoy using. So do not wait for them , go for them.

References

1. **Lancaster, F.W.** "*Future Librarianship : Preparing for an Unconventional Career*". Wilson Library Bulletin , 57, 1983, 747-753.
2. http://www.suite101.com/content/library-20-and-remixing-existing-services-a126105
3. **Wan We Pin.** "*Library 2.0 : The New E-world Order*". Library connect, Vol.5 No. 4 .

16

Print and E-Resources : A Way Towards Hybrid Library

Rabindra nath Mohanta

Abstract

ICT is changing the environment of Libraries and Information Centers in India. The advent of Internet and subsequent services has led to dynamic changes in communication and use patterns in libraries all over the world. Most of the advanced libraries in India are providing traditional as well as ICT based services. This paper defines hybrid library, the need for Collection Development Policy (CDP), allocation of library budget according to the requirement of the users, and the need and services of hybrid library are discussed. The librarians and teachers should work together to motivate the users in the use of E-resources. The paper suggests developing manpower planning for the hybrid library and staff formula to be formulated by UGC in the present era. The LIS centers such as NISCAIR, NASSDOC, and INFLIBNET should take the lead role in conducting manpower training for LIS professionals across the country. The services of UGC/INFONET, FORSA consortium and INDEST consortium services should be taken by the library for the users.

Introduction

ICT is changing the environment of libraries and information centres in India. The advent of INTERNET and subsequent services have revolutionized the concept of libraries and changed the way information is processed, stored, transmitted, retrieved and disseminated. It has large volumes of electronic information in almost all fields of human knowledge. Most of the libraries in India are providing traditional as well as ICT based services. Information is available in abundance from many sources and many formats such as oral, print text, non-book material, television, Videos, Library Data bases, Websites, E-resources such as E-books, E-journals. The hybrid library is on the continuum between the conventional and digital library, where electronic and paper-based information sources are used alongside each other. The challenge associated with the management of the hybrid library is to encourage end-user resource discovery and information use, in a variety of formats and from a number of local and remote sources, in a seamlessly integrated way. The hybrid library should be "designed to bring a range of technologies from different sources together in the context of a working library, and also to begin to explore integrated systems and services in both the electronic and print environments." The hybrid library should not, then, be seen as nothing more than an uneasy transitional phase between the conventional library and digital library but, rather, as a worthwhile model in its own right, which can be usefully developed and improved. Hybrid library is a place in which information can be accessed both in hard copy and electronic formats. The resources of the hybrid library are both electronic and conventional, and their functions are automated. Users may search the resources either online or offline. Use patterns have witnessed a paradigm shift from print to electronic materials because of proliferation of E-resources. E-resources consists of wide verities of materials including books, indexes abstract, Encyclopaedias, Reference Books, Aggregator, Data bases and full text or partially full text data bases. As the resources change at a very rapid pace and as libraries continue to build larger collections of E-resources finding ways to manage them effectively from selection to licensing is becoming a major challenge for the librarian of the 21st Century. Thus, libraries need to adopt both print and electronic resources for its collection development in a better way to fulfil the requirements of users.

Collection Development in Hybrid Library

Collection Development refers to the collection of different types of library materials keeping in view the changing requirements of the users. Collection Development is a planned function. It should be done judiciously. Collection Development Policies decisions are made at the highest level. The library must develop a CDP consisting with the objectives of the institutions and the users of the library. The information sources should satisfy content, currency and quality. Collection Development in any library is not an easy job. This should be based on policies and programmes chalked out of representations of faculty members from various faculties in cooperation with the librarian. The involvement of all staff members of the library is also essential for developing a balanced collection.

Manpower Requirement for Hybrid Library

Manpower is very important component in every library. Man power requirement in present age library i.e. hybrid library should be framed according to new staff formula to be worked out by UGC. Frequent advancement in ICT makes the task of training library staff more challenging. LIS researchers have frequently mentioned the need to improve staff ICT skills and expertise. Man Power Policies should be framed for continuing education and for in-house –training and hand on experience for the LIS professionals to use modern technologies and latest ICT gadgets. However, there is currently no uniform way of managing and providing integrated access to these hybrid resources. Users are forced to interact with each service individually and waste time in repeating the same steps to search different systems. The LIS Schools, NISCAIR,

NASSDOC,NISSART, INFLIBNET should take the lead role in conducting manpower training for LIS professional across the country. Hybrid library should plan for human resource development and impart training, motivation, and also change the mindset and provide facilities to the existing staff members to make them capable of handling E-resources.

Library Budget for Hybrid library

Budget of the library should be judiciously prepared according to need of the users. The allocation for purchase of E–resources and printed sources should be in separate "Heads" in the budget for the hybrid library.

Library Services for Hybrid library

The hybrid library should provide access to the most current reference sources available in order to provide the accuracy and quality of information. E-Awareness programme for the users should be organized at regular interval. The librarians and teachers should work together to motivate the users in the use of E-resources. The staff should also help the users to find the information from different sources. Thus, high quality LIS services can be provided in an effective way.

Library Consortia in Hybrid library

Many Indian university and college libraries are not in a position to subscribe to all the required journals and databases mainly due to lack of management support and financial constraints. The libraries are forming consortia in order to facilitate knowledge sharing at a much cheaper rate. It has been observed that the E-journals consortium in India saves subscription up to 90%. "Publishers offered to the consortium are lowered by 50% to 90 %, lower rates of subscription in India by the publishers are not only because of combined strength of its members, but also due to their eagerness to enter into the Indian market."[3] The services of UGC/INFONET, FORSA consortium and INDEST consortium services should be taken by the library for its users.

Conclusion

The library and information services of the 21st century are fast changing. The services of most of the libraries are not confined within four walls but are integrated into local, regional, national and even international networks. With the rapid development of E-resources libraries should acquire both printed and electronic sources of information because E-resources are the most nascent entity in libraries, considering the long experiences with print media. E-resources are going to continuously increase day by day due to users demand and requirement. Libraries have to cope up with the latest ICT gadgets, E-resources as well as the knowledge of print media to serve the users need and requirement in the present era. Thus, hybrid library is preferable till digital technology finds solution to some of the problems encountered by it at present.

References

1. **Chris Rusbridge.** "*Towards the hybrid library*" D-Lib Magazine, July/August 1998, URL: http://www.dlib.org/dlib/july98/rusbridge/07rusbridge.html (accessed on 5th April 2011.).
2. **Pradhan, Debasis and Tripathi, Tridid.** Digital Library, Virtual Libraryand Other Emerging Library Systems. *IASLIC Bulletin*.54,3;2009.p. 157-162.
3. **Shah, P,C.** Editorial *Dlibcom* .4,10;Oct2009;p.2
4. *Dynamics of E-resources & Usage Trends in Digital Era* edited by **R.K. Mahapatra and others.** New Delhi : SSDN Publishers and Distributors, 2011.

17

The Impact of Information and Communication Technology in Library and Information Science : E-Publishing, Opportunity or Challenge

Rajesh Kumar Maurya &
Anand Kumar Tripathi

Abstract

As world is running with new trends, a new era of E- publishing has started to cope up the demand of knowledge at just one click. Electronic publishing is basically a form of publishing in which books, journals and magazines are converted into digital form. These publications have all qualities of the normal publishing like the use of coolers, graphics, and images and are much convenient also.

Electronic publishing empowers all writers in ways that no technology has ever done before. The use of licensed electronic information resources will continue to expand and in some cases become the sole or dominant means of access to content without violating any issues of privacy or Confidentiality we can dramatically enhance our understanding of information use. Digitization is not only beneficial for the environment but also very challenging in its own aspect. This paper is to discuss the advantages as well as challenges in digitization.

Keywords: *Electronic publishing, ICT.*

Introduction

The impact of information technology has been all pervading. It has changed the way that information is stored and disseminated and has threatened the traditional approaches to the library and its services. The ultimate objective of any information system is utilization and exploitation of information that raises levels of education, strengthen community links and stimulate participation in decision making for development of mankind. Information is an intellectual resource that has the capacity to change the image of recipient. In the postindustrial society, it has been said that what counts is not row muscle power or energy but information. Consequently, large investments are being made in the information technology industry for the purpose of generating, processing and disseminating information.

The information produced is disseminated by different model of publishing information. New technologies have transformed the process of publishing and distribution of information. In view of growth information, electronic publishing has become a foundation for the new information society to get the right information to the right person at the right time. Changes in the publishing industry have a direct impact on the information systems and services. This information technology has altered the mode of publication in such a way that though the traditional sources of information continued to be flooded with the attractive electronic form of publications. In the changing scenario, libraries and librarians will have to play a crucial role in handling conventional and electronic resources.

Thus the era of electronic publishing has begun affecting producers, distributors, and library and information professionals. The ultimate goal of electronic publishing is to provide fast and easy access to the information contained in the objective publications with simple, powerful search and retrieval capabilities. Thus, e publishing can be used effectively in the context of Dr.

S.R. Ranganathan's fourth law "Save the time of user" for many purposes.

Applications of E Publishing

What started out as a trend to provide printed material in a digital format, has today become an alternative in itself.

While e- publishing has existed for over 7 years, debate over its use and acceptance has reached a fever pitch in the last few years.

Where can it Find Use?

Simply put, almost everywhere. It is most ideally suited for publications like journals, research reports or resource bases and newsletters. It is also suited for all information that is dynamic or constantly changing, it could be ongoing research or even news.

It finds great use and acceptance in academics, in the online publishing of educational books ǒr tutorials. With an increase in distance learning programs, the need for quality educational material is on the rise. These e-books and study material need to recreate an active learning atmosphere as can be found in a class full of students and a teacher. The content and the product therefore, have to be interactive. This can be done through open assignments, online debates, chats and discussion forums. These products also have to provide for a feedback mechanism. Naturally, in order to be effective these products have to be more colorful and interactive, and at the same time also be concise and also crisp to match the attention span of the student reading online. Electronic-Publishing is also used extensively in large academic institutions and libraries for their periodicals and journals.

E Publishing could also help in bringing out industry journals/ periodicals, rezones and newsletters. Authors and publishers gain when they publish e-books. Costs are lower and royalties could be higher. Working papers on ongoing research helps unleash the power of collaborative work through networking. E publishing is very effective here as different sets of people can work with data so varied and complex, with charts, tables and images of amazing complexity made accessible in digital formats replete with animation and sound almost instantaneously.

Types of E-Publishing Models

There are several models in e-publishing ranging from commercial e publishing, print-on-demand and subsidy e publishers, no fee distributors and self publishers. In addition to these, there are agent vendors who package on-line journals, books and sell the database to industry bodies, universities etc, with a mark-up over their cost, incurred for procuring rights. These costs would vary depending on whether the vendor has procured exclusive rights or has to share them with a competitor. As a variation of the above, there are online bookstores.

Commercial e-publishers are almost like their traditional counterparts, they choose to publish books that are most likely to be sold–a good mix of quality and subject matter is what they look for and rejection rates are pretty high. While their websites serve as effective storefronts, they also sell through other online bookstores like Amazon.com etc. By and large, e-publishers do not pay an advance, but royalty payments are higher than what traditional publishers offer, often as high as 40%. The editorial and technical work that the e publishing team puts in is no different that what is done in print publishing.

Subsidy e-publishers differ from commercial publishers in several ways. For one, they accept most manuscripts and publish them on an "as is" basis. These are not edited or proofread or even formatted. They have a virtual rate card for add-on charges for these services, should the author opt for it. Everything, from adding images or graphics to copyright registration, is offered at add-on rates. They publish books for a flat fee and pay the authors a royalty that is comparable to what they would receive from the commercial e-publishers. However, they are more likely to sell through online stores than through their own websites. Likewise, they pass on the added cost of selling through online stores to the author who receives higher royalties for books sold through the publisher's website and lower royalties if they were sold through, say, Amazon.com.

Another interesting model is the "print-on-demand" (POD) model, which is a mix of electronic and print publishing. The book is held by the publisher in electronic form and is printed out in the hard copy form only on order. This is also done for a flat fee. While many commercial publishers also offer the POD format, this is a model more often followed by the subsidy e-publishers.

The No-fee "Distributors" accepts manuscripts and publishes them on an "as is" basis, in a sense, simply providing these authors with a bookstore. They do not charge a fee, but take a slice of the royalties. Price setting is usually the author's prerogative. The no-fee distributors do not offer services like editing or formatting. This works well for the authors who prefer the self publishing model, but want to offload the task of setting up their own "store front" or website. This also does away with the need to register as a retail business as they get paid royalties by the distributors as opposed to revenues on sales.

As an alternative, several authors prefer to self publish their books. This is not only inexpensive, but also offers them total control over the various processes. The author usually does all promotional and marketing work on the Web. Unless the author wants to offer the e-book in multiple formats like downloads, CD-ROMs or floppy disks, distribution costs are.

Electronic Publishing can be Represented as

Electronic Publishing = Electronic Technology + Computer Technology + Communication Technology + Publishing.

One of the most complete definitions of Electronic Publishing appears in a popular electronic encyclopedia (Grolier Electronic Publishing, 1995). This wholly electronic publication defines electronic publishing this way "the term E-publishing refers more precisely to the storage and retrieval of information through electronic communications media. It can employ a variety of formats and technologies, some already in widespread use by businesses and general consumers and others still being developed. E-publishing technology can be classified into two general categories:

- Those in which information is stored in a centralized computer source and delivered to the users by a telecommunications systems , including online database services and videotext represents the most active area in E publishing today", and
- Those in which the data is digitally stored on a disk or other physically deliverable medium.

Models of Electronic Publishing

E- Books

The book is quiet popular document to meet the academic and general needs of user community. Project Gutenberg, perhaps the best know publisher of book length electronic texts, began in 1971with the goal of encouraging the creation and unlimited distribution of 10,000 electronic texts by the end of the year 2001. These books fall into several main categories including light literature, the classics, reference books, and other literary fields.

Publishing a book electronically is to achieve quick publishing and dissemination of information. A book may not have contemporary value that a journal has but it certainly has an archival and reference value. A number of encyclopedias do come out on CD-ROM. It is felt that the Internet is not a satisfactory platform for publishing full text of documents but CD-ROM is appropriate medium for publishing books. Book length E-text is also available on Floppy disc and CD-ROM, although distribution by floppy disc is decreasing due to the convenience and growing popularity of CD-Rom. Most E-Texts published on CD-ROM are public domain works including encyclopedias. Using the E-publishing language on Internet like SGML presented and published attractively with multimedia effect especially for documents like Yearbooks, Encyclopedias.

Electronic Periodicals

This new media is a vehicle of scientific communication and purely a product of scientific research. This category includes electronic journals, newsletters, magazines, zincs and discussion lists. Perhaps no other area in E-publishing has received more study than the area of E-journals, particularly as they apply to scholarly research.

Electronic Database

With the emergence of computers & communication technologies the strength of information system in the development of modern database has taken new shape. Information originating from a database has become a large segment of Electronic publishing that provides a base for procedures such as retrieving information, drawing conclusions, and making decisions. The holding of the library database consisting of books, periodicals, reports & theses can be converted to electronic form that allows access for public use through digital networks. The online electronic library card catalog (OPAC) shows how information could be published and that enable user to search the document with various access points like author, title, subjects. The six major developments that have taken place since the 1960s that have squared the use of databases are: -

1) Networking and co-operative arrangements.
2) Leasing arrangements for information access.
3) Establishment of scientific information dissemination centers.
4) Increased online access (remote access via terminals).
5) Emergence of the online retails (i.e. Dialog, Lexus etc).
6) Improved distribution (via CD-ROM or other means).

Various electronic databases publishers today account for publishing information both bibliographic and full text on CD-ROMs as well as making them available for online retrieval. The prominent online publishers include DIALOG, BRS, and EBSCO host etc.

An excellent example of electronically published databases, the ERIC (Educational Resource Information Centre) database is the largest educational database in the world that contains more than 800,000 records per year. ERIC is available in CD-ROM format as well as on the net free of charge.

Electronic Ink

Electronic Ink is a developing technology that could have a huge impact on the media and publishing industries. Electronic Ink could be used to create a newspaper or book that updates itself. In addition, this content could be programmed to change at any time. For example, you could have a billboard that rotates different ads, or you could receive a coupon in the mail that is frequently updated with the latest offer. For media companies, the possibilities are almost endless. Someday your electronic newspaper will simply update itself every day. E- Ink Corporation, a new company with major investors, and Xerox are two companies currently developing this technology.

Electronic Publishing on CD-ROM

CD-ROM has provided new dimension for information storage and retrieval. Publishing information mainly abstracting sources are quiet common in CD-ROM. Although much of the work on e-journals has concentrated on distribution via the Internet, there has been some work on CD-ROM as well. The advantages of CD-Rom are-

- More material can be included, both in terms of quantity (650+megabytes) and type (multimedia.
- Full text searching is relatively easy to include.

Considerable savings in cost a 1000 page book may cost as much as $14 per copy to produce in quantities under 1000 while many replication plants will produce 1000 copies of a CD for less than $2 each.

Digital Content

Digital Content generally refers to the electronic delivery of fiction that is shorter than book-length, nonfiction, and documents and other written works of shorter length. Publishers of digital content deliver shorter sized works to the consumer via download to handheld and other wireless devices. Technology used for delivering digital content includes Adobe PDF, XML, HDML, WAP (Wireless Application Protocol) and other technologies. The security of the data being delivered is the major concern of publishers who want to ensure they can deliver digital content without the risk of someone copying the work and selling or giving away the works.

Email Publishing

Email publishing, or newsletter publishing, is a popular choice among readers who enjoy the ease of receiving news items, articles and short newsletters in their email box. The ease of delivery and production of email newsletters has led to the development of a massive number of available email newsletters, mailing lists and discussion lists on a large variety of topics. Newsletters are also widely used by media companies to complement their web and print offerings. Many authors and writers publish their own newsletters in order to attract new readers and to inform their fans about new books and book signings.

Web Publishing

Web publishing is not a novel practice any longer, but it continues to change and develop with the introduction of new programming languages. HTML is still the most widely used web programming language, but XML is also making headway. XML is valuable because it allows publishers to create content and data that is portable to other devices. Nearly every company in the World has some type of website, and most media companies provide a large amount of web based content.

Major Challenges and Opportunities

All LIS schools recognize the importance of infusing more ICTs in their curricular. But their wishes and efforts are constantly thwarted by a number of key challenges. These include:

Challenges

Inadequate technological infrastructure to support the integration of ICTs in the curricula. This refers to issues as poor or lack of national ICT policy, low internet connectivity, inadequate supply of electricity, inadequate number of PCs, etc. There is need for policies that deregulate satellite communication and other telecommunication links, regulate ISPs, regulate government and cross-border data flows, etc. ICT policies can help address stringent tax regimes that still treat computers, communication equipment and other peripherals as luxury items, thus imposing heavy import duties on them and subsequently rendering these items very expensive.

Internet access is now widely available, but the efficiency is poor as many LIS schools experience downtime, several times a week. The telecommunication services are the root cause of these downtimes in terms of, either, low bandwidth, technical faults and other network configuration problems. As Jensen (2005) puts it, there are also "many external systemic factors such as electricity, transport networks, import duties" etc, which impact on internet service delivery on the African continent. In some institutions, access is limited, not only by the number of Internet service points, but also by the time that access is available or permitted, leave alone the difficulty of bandwidth. Yet for research purposes, access to the Internet is no longer a luxury or privilege for only a few people because in academic circles, access to the Internet and hence to the world's stores of knowledge is a necessity. LIS departments still need to lobby to gain greater access to Internet resources for academic staff and/or research. Thus there is urgent need for improved ICT policies and infrastructure in institutions and countries

Funding/sustainability of the technology is the major non-technical constraint in LIS schools (Minishi-Majanja, 2004). Most universities decry the issue of under-funding in most of its functions. Besides, the unprecedented, phenomenal and multifaceted growth and development of the ICTs themselves pose another challenge. This rapid pace and transient nature of technological development requires sustained funding. While the centralization of ICT services, hence funding, has been found to be the most affordable system for institution-wide development and use of ICTs, it only works well where there exists a policy that has explicitly incorporated the goals and needs of all sectors, including those of the LIS school. In institutions where the political economy is slanted, coupled with the absence of such a policy, a LIS school may suffer from neglect and hence be unable to develop and use ICTs.

Expertise among the constraints cited by respondents in the study were the issues of (*a*) re-skilling lecturing staff so as to improve their ICT competency, (*b*) lack of systems manager/support staff and/or ICT experts, and (*c*) low levels of students' epistemological access (Minishi-Majanja, 2004:227). Manda (2006) observes the lack of ICT knowledge and skills among staff. Ikoja-Odongo (2006) decries the problem of brain drain i.e. that staff sent overseas for training either do not return to their posts or are taken up by other organizations that are able to offer them higher remuneration. This suggests that in so far as re-skilling academic staff is concerned, opportunities are available but there is still no guarantee that the problem will be solved because of the prevalence of skills shortage at macro/national levels.

There is still a serious need for technical support staff with high level expertise in the maintenance aspects of ICTs. Because of poor maintenance and insufficient skills to diagnose system problems and swap parts, there are many out-of-commission machines which could easily be re-activated and used. The problem of technical expertise is two faceted. In the first place, there are not enough people qualifying or attaining ICT specialist skills at the speed at which the technologies are adopted. Secondly, the problem of brain-drain whereby the few experts opt for better paying jobs overseas.

Ikoja-Odongo (2006) observes that many students join the university without any computer skills and hence much time is taken trying to make them computer literate. However, it seems that the problem of students' ICT skills may be short-lived because on the one hand, computer literacy has now been introduced at many school and colleges, while on the other hand, the 'N' and 'Y' generations are becoming of age.

Opportunities

Expanded job market. The market demand for LIS graduates who have strong ICT skills and broad perspective on information management has expanded. Both the public and private sector have recognized the importance of effective management of their knowledge and information resources. However, many of the organizations in these sectors do not necessarily want the traditional LIS perspective. Rather, they need a versatile professional who is able to actively participate in detecting cues for relevant information, gaining/providing access to relevant information sources, searching and synthesizing data, repackaging information, and adding any other value that enhances the effectiveness of the organization. All these need extensive ICT knowledge and skills that LIS education can effectively integrated in their curricula.

African LIS schools should continually review their curricula and innovatively infuse a stronger ICT component. Apart from the established procedures of curriculum review, heads of LIS schools need to network and keep in contact with colleagues in other LIS schools through correspondence, email or conference attendance so as to pick up new ideas. Additionally, LIS schools should explore methods of collaborating with their counterparts (such as other LIS schools, academic staff or non-LIS departments), who have expertise and resources to offer modules/competencies that they may be unable to offer. Where academic staff expertise is the problem, but ICT facilities are available, this can be achieved through distance education and/or online education.

It is encouraging to note that some international companies, such as Bull, Compaq, IBM, NCR, Oracle and Microsoft operate offices in Africa with reliable local representation in most countries. The presence of these companies and/or their representatives indicates their appreciation of the market for their products, and it would be in their best interest to maximize their exploitation of the market rather than concentrate only on the corporate Africa. In any case, PC equipment is often clone equipment imported from Asia, but Compaq, Dell, IBM and ICL also have significant shares of the market and Dell South Africa is now selling via the Web

(Jensen, 2005). There is also the growing availability of high-speed wireless Internet access, which is often seen as Africa's information revolution. As Van deer Merowe (2003) observes, "Africa needs this system because it makes the continent more accessible to international business people and business... But more than that, the wireless technology lends itself to the rapid, low-cost roll-out of a new wave of connectivity into rural areas not currently served by telecoms" Jensen (2006) observes that countries can strategically improve infrastructure policies that can minimize bandwidth problems, for instance, by obtaining "access to national and international backbones at cost, rather than at the high tariffs charged at monopoly prices by the incumbent operators." Nevertheless, LIS schools in sub-Saharan Africa should continually urge their parent institutions to draw realistic budgets and provide sustained funding for the ICT projects. Individual institutions and departments must try to find ways of obtaining the necessary funds, be it through income generation activities or liaisons with the private sector. Additionally, some cost-cutting measures can also be employed, such as those suggested by James (2001), i.e. the use of open source software or cheaper versions of software e.g. New Deal, Office2000, etc. which can also operate on older hardware; procurement of refurbished computers distributed by such organizations as New Deal, Freedom, Computer Aid International, and World Computer Exchange; redesigning of hardware so as to lower the cost of Internet access, for instance using hardware that does not have hard drive or disc drive but has Internet software; merging internet technology to use television connection with modifications; and using community wireless LANs e.g. Air Port (http://www.freebase.sourceforge.net).

Conclusion

We are living in an Information society demand for information is growing constantly and its value escalates in direct relation to that demand since the explosion in recent years of the World Wide Web and other electronic resources onto the information, the response is positively overwhelming by those who work with information resources. Electronic publishing or e publishing uses new technology to deliver book and other content to students and readers, the technology allows publishers to get information to readers efficiently and quickly. It has become very popular technology in Information system today. A Library and Information Manager should familiarize himself with the emerging technologies mentioned above This technology is developing at a very fast rate and what looked impossible a few years back is becoming a reality now. The only thing we can be certain of is that there will be radical changes and in some instances there will be rapid. Thus by the turn of this century the whole scenario of library operation will occupy an important place in the 21st century.

References

1. **Grater, D.W.**, "*Electronic Publishing: Evaluation, Procurement & Management*". Tab Professional& Reference Books,1989.
2. **Chandrakar, Rajesh.** "*Electronic Publishing Model for Indian Academic Journals*". Presented at International Conference on Digital Libraries, TERI, New Delhi, 5th-8th December 2006.
3. **Stander, Old rich,** "*The Electronic Era of Publishing : An Overview of Concepts, Technologies & Methods*", Elsevier, 1987,p.189-203.
4. **Mastroddi, Franco,** Electronic Publishing, "*The New Way to Communicate*", Kogan Page, 1987, p.338-352.
5. **Eysenbach, Gunther,** "*The Impact of Preprint Servers and Electronic Publishing on Biomedical Research*", Electronic Publishing, December 1993, Vol. 6(4), pp. 499-502.
6. **Delamothe, T., Smith R., Keller M.A., Sack J., Witscher, B.,** "*Netprints : The Next Phase in the Evolution*" of Biomedical Publishing, Brit Med J 1999, 319: 1515-1516.

18

Print as Well as Electronic Resources: How Much is it Essentials for an Academic Libraries

A.K. Rai, Santosh Kumar &
Amitabh Gupta

Abstract

Libraries have supported multiple formats for decades, from paper and microforms to audiovisual tapes and CDs. However, the newest medium, digital transmission has presented a wider scope of challenges and caused library patrons to question the established and recognized multi format library. Within the many questions posed, two distinct ones echo repeatedly. The first doubts the need to sustain print in an increasingly digital world, and the second warns of the dangers of relying on a still-developing technology. This article examines both of these positions and concludes that abandoning either format would translate into a failure of service to patrons, both present and future.

Introduction

With the advent of the Web and the proliferation of electronic information, librarians are more frequently confronted with questions from their administrators and patrons on the present and future value of the printed book. Technology and its advantages—convenience, cost, timeliness—present an appealing future. So why libraries would continue to stock their shelves with printed texts, and why should their parent institutions provide space or funding for such acquisitions?

The answer is surprisingly complex, but it begins with the library's mission to meet current and future user needs. Not only does print still dominate in scholarly publication, but the rapidly changing technological environment contains inherent future risks for researchers and academic libraries. The same creativeness that spawns advancement and development also slows or prevents the adoption of stable standards in format, laws and pricing. An environment that blossomed under constant change does not react well to restriction, and tying research or development to a particular standard would result in stagnation. Without such standards, though, a library cannot guarantee the preservation of information for future generations, and may even be frustrated in its goal to provide continued access to the current generation.

Some university administrators will recognize this refrain and will join in demanding, "With such unpredictability, why collect electronic resources at all?" A partial answer resides in the convenience mentioned earlier, but it is only one facet of the analysis. The world of information gathering is in transition, and users have shown a marked preference and reliance on a technology, which, if not stable, will undoubtedly continue to thrive. Libraries, as information brokers, cannot reject data simply because it fails to comply with existing expectations or because its format of transmission is not yet fully developed. Instead, they must seek to harness its strengths and to educate users on its weaknesses.

The librarian's responses to the administration's questions on both sides of this debate have wide-reaching impact, from space allocation and budgeting to the content of the library's permanent collections and general user education. This article is designed to provide data to support the proposition that a twenty-first-century academic library requires *both* traditional print materials and electronic resources.

Print Values

The core purpose of an academic library is to serve the needs not only of today's users but also tomorrow's. It follows that the library must have an enduring collection of resources that is accessible and meaningful to both current and future scholars. For the reasons detailed in the following sections, physical formats will remain in collections for many years to come. As the world has seen with radio after television's debut, the creation of a new medium does not necessarily invalidate the former ones. Print is a time-tested format that continues to fulfill promises that technology cannot yet deliver.

Why Online

A common fallacy is that all information is available on the Internet, whether free or through a fee-based service. Despite tremendous strides in electronic publishing and in digitization technologies, the majority of the world's published materials remain in physical (print or microform) formats only. Since the invention of the Gutenberg press, the publishing world has produced more than five centuries' worth of materials, and recent efforts to digitize scholarly historical publications2 have covered only a small proportion of these. For colleges and universities, this historical component is particularly worth noting, as much of knowledge research depends on analyzing knowledge's evolution.

Even in recent technology-friendly years, the printed book has not been made obsolete by e-publishing. In fact, in 2010, the output of print publishers outpaced that of previous years, and most of the produced titles were not available in electronic Format Notably for academic libraries, the majority useful materials are not attainable in e-book format. These numbers demonstrate that libraries seeking to serve their patrons must continue to examine, evaluate, and collect print materials.

Electronic Material (Not Always Free)

A law library, even the well-endowed one, has a finite budget, and its resources must be judiciously allocated among a range of interests and needs. Users who do not directly authorize purchases frequently overlook the cost component involved in selecting a library resource. With so much information available to Web surfers, it appears as if no-cost is the norm for e-resources. However, libraries can and do arrange for seamless authentication to fee-based resources, thereby eliminating the requirement for individual user passwords and identification information; thus, a resource appears "free" to end users. Such users are often unaware of any restriction to the resource until they attempt direct remote access and are prompted for a password.

Of course, it must be acknowledged that print resources are also costly, and administrators may ask why expense is a differentiating factor in format selection. The answer is twofold. First, with tangible objects, there is rarely an assumption that no cost was attached to its acquisition. The "free" assumption in e-resource access originates from the intangible nature of electronic information. Second, the discussion later on licensing will illustrate why pricing for online resources may have a greater impact on libraries and their patrons than pricing for physical formats.

Electronic Material (Not Always Accurate or Authenticated)

Studies have cited the growing reliance on Internet research by the public for widely ranging issues from health to current awareness. For an academic library, though, in which patrons require authoritative data to support arguments, instruction, or scholarship, the greatest strength of the Internet also becomes one of its dominant liabilities. Anyone with a computer and the necessary rights to a Web server can post or alter data. This greatly increases accessibility and availability of information on a limitless range of topics, but it also means that anyone with such access can edit documents and disseminate false information, actions that cannot necessarily be detected by the user. There is no guarantee of the poster's authority or of the authenticity of a document. Some assurance and security derives from a reputable site or designated domain (e.g., .gov or .edu), but even in such cases, the review process for documents posted is not always apparent or consistent. Unlike most printed resources (except vanity publications), many free online ones are not routinely reviewed, edited, or checked for accuracy prior to or after publication. Even fee-based e-resources may inaccurately report content information, from titles or dates included to currency.

Print materials are also subject to mutilation or defacement, but in such instances the original text itself is unchanged; only the copy that the individual library holds is damaged. With e-resources, the source document can be modified as easily as any copy, without indication as to when the change was made, who authorized it and what motivated the alteration. Destruction of print materials results in the removal of an individual volume from a collection; destruction of an e-resource corrupts the source document, making verification or retrieval of the document by later generations impossible.

From the academic perspectives, this unreliability fostered an early distrust of the e-document by scholars and tenure review committees, which in turn contributed to the dependence on and expansion of print publications. Acceptance has since grown in both arenas; court opinions increasingly refer to Web sources and tenure review committees will consider reputable online journal articles when accompanied by proof of stringent peer-review. E-books/monographs, however, have not been embraced by the scholarly publisher, and in this area, the printed text remains the standard.

Stability

Having endured centuries of societal and mechanical meddling, the manner and result of physical printing is firmly established. Few uncertainties surround the printing or acquisition processes, user control of the printed form, or the standards to be used in publishing. In contrast, the transitory nature of technology renders access to digital documents unpredictable.

First, all sites, including fee-based ones, support the addition and removal of information without the consent of its users, while such freedom is necessary to keep information up-to-date, it also facilitates the unannounced removal of useful documents.

Second, no set standards are uniformly accepted and applied to e-documents. Without standardization, multiple proprietary and open access formats are created, supported, and distributed to users. As technology marches forward, the lack of a standard will make transitions to new formats burdensome and will likewise impact accessibility.

Permanence and Completeness

Many electronic documents are fleeting, ephemeral. They can disappear entirely or suffer from the lesser defect of link rot. They may be removed from databases by the publisher for reasons ranging from desire to keep the database current, to disinterest in maintaining a low-use resource, to fear of litigation over Database Copyright issues, to objections of a political nature. Alternatively, electronic publications may cease to be updated or may move without notice.

This unpredictability increases when the data provider differs from the original publisher. Not only does continued access to the resource depend on a contractual arrangement between the publisher and the provider, but the provider also has the right to selectively post the information licensed. Such alteration manifests itself in a variety of forms, including the exclusion of charts and pictures, selective publication of articles, and incorrectly described scope notes.

If the drafts are published in print, libraries can obtain them immediately through purchase or later through interlibrary loan. When drafts are born digital, however, publisher disinterest or desire to provide only the most current information may compel it to overwrite the draft document(s). Only a few publishers, like GPO Access, consciously and actively preserve electronic records of deliberations and drafts. Uncertainties over the preservation of such electronic drafts often lead libraries and users to print out important documents for future use, and such printing results in later questions about authenticity and validity.

Standards and Preservation

"Changes in computing technology will insure that, over relatively short timeframes, both the media and the technical format of old digital materials will become unusable. Keeping digital resources accessible for use by future generations will require conscious effort and continual investment."

Unlike print, electronic documents are not immediately readable by the naked human eye. The ability to access them is dependent on equipment and technology, both of which become important considerations when examining the acquisition of information stored in electronic form. Vinyl records, 8-track tapes, betamax cartridges, and 5-¼-inch floppy disks have aptly demonstrated that reliance on any one technology, especially intermediate ones, portends significant future expenditures. To keep the data on such formats accessible, information must be migrated to new formats. Barely ten years have passed since the World Wide Web, as currently known, came into being. In that time, there have been multiple changes in "markup languages" (e.g., HTML, XML) used to prepare documents for Web publication and in information format choices (e.g., JPEG, TIFF, PDF). In evaluating a resource, libraries must consider whether the data is of permanent or only temporary value to the collection; if permanent, it must factor in future costs for migration or preservation.

Print texts have lifespan counted by centuries, determined by paper quality and the elements that touch it, and extended by the application of ANSI/NISO standards and preservation techniques. E-documents' lives are briefer, due in part to the absence of set standards in digital production. According to the

National Archives, the United States federal government alone uses at least 4800 different formats for electronic records; others place the number at 16,000. Conversion or preservation of these documents presents a staggering obstacle for the government and other e-document providers, while continued access to physical materials is not similarly questioned.

Archiving

The problem of preserving digital information cannot be solved definitively, at least not as long as information and communication technologies continue to change, because such change alters the character of the problem.

The printed word, once printed, is fixed and unchangeable. The content of a book is archived as soon as the book is printed, since it is a stable, self-preserving format that endures for centuries. In contrast, electronic documents are not self perpetuating. Unless contents are actively backed up, consistently converted to current technologies, or both, an e-resource may not be accessible by tomorrow's user. Until standards are reached on issues such as archival responsibility and methodology and format preservation, electronic-only collections jeopardize a library's long-term obligations to build and maintain a usable collection.

Further, Web sites and information stored thereon change constantly. Even if standards on archiving and preservation are universally adopted on specific Web-based documents or databases, much of the information previously published on the Web, including most of that available today will not be saved. A news article on a single site can change multiple times in a day. Electronic modifications are not easily identifiable, and patrons seeking to find a particular version of a document may be unable to do so. Though ventures like the Way back Machine (www.archive.org/web/web.php) are archiving selective free pages and materials, they cannot save every version of every page on the Web. They are also unable to access fee-based resources, thus leaving extraordinary amounts of valuable information to disappear without a trace. The digital format should be harder than print, since it does not degrade over time. Ironically, due to its incidental needs, digital documents are being archived in the more reliable physical format (print or microform).

Smart Use

The evolution of e-documents has just begun, as is most evident for the struggling e-book reader. While reference materials, indexes, and news clips lend themselves well to the digital medium, studies have shown that technology has not been able to replicate the ease of book reading. Over the ages, print publishers have optimized font and appearance for readers, and any reduction in this ease is resisted. Also, readers have learned to recognize a book's chapters in relation to each other. In a printed text, they can easily find a particular reference by this tactile and cognitive association. A data file, which is without physical form, cannot support that same type of reference. A new manner of association will develop as the e-book matures, but its present format is not yet ideal (or even very useful) for knowledge research, which relies heavily on such associations.

The dissatisfactions with lengthy digital documents vary from unfamiliarity to difficulty of use. Though e-books have made great strides, moving away from requiring specific reading devices or employing exclusive use technology, readers continue to express a preference for the printed text when any amount of extended reading is necessary. One recent example is the 9/11 Commission Report though posted immediately on the Internet; readers purchased enough hard copies to make the title a national bestseller for more than nine weeks.

Filtering of Exactness

The Internet is free and fair, open and accessible, personal, postmodern, empowering. That old-fashioned phrase, 'the world is my oyster,' is embodied on the Internet. But oyster divers can drown—too much water, too deep, bad currents.

For years, people have averred that the Web's limitless capacity for information storage would displace the printed text. Even as users found more information available to them, though, they found that endless and unfiltered data complicated research by overwhelming them with irrelevant hits or requiring them to acquire resource-specific search strategies. In this analysis, the value of print has only been elevated.

Due to the expense of editing, publishing, and distributing books, printed materials are thoroughly vetted. Manuscript submissions are reviewed for accuracy, uniqueness, and readability. Therefore, a user who picks up a printed text in the library has the assurance that some evaluation of the resource's worth has already occurred, both by the publisher and by the library's selectors.

Online, even where relevance ranking is available, the worth of a document and the elements used in determining worth are not always easily discernable. Free resources are particularly troublesome in this regard. Engines may allocate greater weight to their advertisers' sites, relevance may depend on meta tags or other self reported information by site owners, and results may depend on the intentions of other searchers, as in the case of popularity-determined search engines. The Web is not designed to be exact; not all pages are searchable by the usual engines and not all pages are equal.

Equally important, few users realize that most search engines index less than 10% of the Web. Only the static pages are easily searched, while the dynamic ones (e.g., directories) are frequently excluded. Search engine users may mistakenly believe that they have searched all Web-posted data when, in fact, they have only scoured a small percentage of it. Even with fee-based resources, which contain edited information and have strictly defined search structures, a user must be aware of the structure in order to search efficiently. Therefore, not only does online searching proffer documents of questionable worth, but their limited capabilities and the nondisclosure of engine biases may mislead users in resource evaluation.

Conclusion

"The mere existence of information does not guarantee its actual use; the format of its presentation has a material bearing on making content either easy or difficult to use". Though Mann's intent was to bolster the use of print materials, his sentiment applies equally to digital data and, indeed, to any newly introduced format.

Recent patron demands for the dissolution of the physical library and its collection are driven by the heady effects of instantaneous gratification provided by technology, the resulting belief that everything can be found online, convenience, and sophistication of use. Undoubtedly, advances in technology will continue to open new doors for researchers, but the more appropriate focus may be the value of any door to information, and not just novel ones. As Mann also states, "Electronic formats are here to stay. Whether they should be regarded as additions to book collections or substitutes for them in research libraries is the point at issue."

Print formats have independent value and contain centuries of information not yet available in other formats. Additionally, the wonders of technology arrive with countervailing questions about preservation, long-term research needs, content quality, document control and authenticity. Technology's nature makes it vulnerable to attack, modification, and disappearance, and its evolution has not yet reached a point where it rivals print in stability, longevity, and ease and comfort of use.

Presently, print and e-formats each have exclusive values, and until those values can be replicated in other media, both formats must be collected, maintained, and supported by libraries. Libraries serve as gateways, and librarians as experienced and knowledgeable guides in the use of emerging and existing media in the pursuit of information. Neither print nor digital information can be ignored or avoided, as both play critical roles in the academic library's survival.

References

1. **Lakshmana Moorthy, A. and Karisiddappa, C.R.** "*Copyright and Electronic Information : in Access Electronic Information*". Paper Presented at the Sixteenth Annual Convention and Conferences on Access to Electronic Information. 25-29 January 1997, Bhubaneshwar, M Mahapatra et al (Eds). New Delhi, SIS, 1997, pp.403-417.
2. *The Hindustan Times*, 23rd September, 1999 p.15
3. *The Indian Express*, 1st November 1999 p.9, 5th November 1999 p.11.

19

Preservation of Electronic Information Resources

Anshu Bansal

Abstract

"The vast amounts of information produced in the world are now for a large part electronic. The rapid growth in the creation and dissemination of electronic information has emphasized the digital environment's speed and ease of dissemination with little regard for its long-term preservation and access. To some extent, electronic libraries, that are those libraries that are moving toward provision of materials in electronic form, have been swept up in this attitude as well. Electronic information includes a variety of object types such as electronic journals, e-books, databases, data sets, reference works, and web sites, which are born digital or which have their primary version in digital form. The advent of electronic information introduces new preservation requirements. In contrast with print materials, where to preserve the artifact is to preserve the information contained in it, electronic information is easily transferred from one medium to another with no loss. The task of managing the ever-increasing electronic objects throughout their life cycle

and to preserve them in perpetuity becomes more and more complex because of they are fragile, volatile and ephemeral in nature. Their viability depends on technologies that are rapidly and continuously changing. This paper deals with the preservation of information in the electronic environment. It also discusses the practical considerations necessary to create a digital archive. This article would help Library & Information Professionals (LIP) to navigate this complex new field."

Key words: *Digital Preservation, Metadata and Preserved Formats.*

Introduction

Major activities have been underway in digital archiving and preservation since the early 1990s. The vast amounts of information produced in the world are now for a large part electronic and include a wide variety of materials: text, databases, audio, film, images. They range from medical records to movie DVDs, from satellite surveillance data to websites presenting multimedia art, from data on consumer behavior collected by supermarket tills to a scientific database documenting the human genome, from news group archives to museum catalogues. The problem on preserving digital records for long term access requires careful consideration about process and technology. Until today there has been no storage platform that can be trusted to store critical electronic record for long time. Preserving digital information is more difficult than preserving record on materials such as paper. The sheer volume and the volatility introduced by digital demand new software architecture capable of scaling and of preventing accidental changes to the records. Procedures need to be put in place to identity, classify, move, evolve, access and occasionally dispose of digital records. Library and Information science and traditional archival practice provide an extensive body of knowledge that can be leveraged with technology create a true modern archive.

What is to be Done?

What must libraries do to address the challenges of digital preservation? There are a number of key areas in which concerned institutions and professionals can contribute. These are:

- Production and Creation of electronic Information
- Ingest: Acquisition and collection development
- Data Management
- Formats for preservation
- Digital Triage: Developing Guidelines for what can and should be saved
- Rescue Operations: Ensure Vital Electronic Documents are Preserved
- Being Legal
- Promoting the importance of Preservation
- Working Together
- Digital Preservation as Public Good

Now, we will discuss about these points.

Production and Creation of Electronic Information

Preservation and permanent access begin outside the purview of the archive with the producer or the creator of the electronic resource. This is where long-term archiving and preservation must begin. Information that is born digital may be lost if the producer is unaware of the importance of preservation, and practices used when electronic information is produced will impact the ease with which the information can be digitally archived and preserved. The archiving and preservation process is more efficient when attention is paid to issues of consistency, format, standardization and metadata description before the material is considered for archiving. Limiting the variability of the incoming 267 material is easier for a small institution or a single company to enforce than for a national archive or library, where a variety of formats must be ingested, managed and preserved. In the case of more formally published materials, such as electronic journals, efforts are underway to determine standards that will facilitate archiving, long-term preservation and permanent access. Such Standardization is considered key to efficient archiving and preservation of electronic journals by third-party archives. Creators can also create metadata at the producer stage, where an expert can support the description of the technical content. With recent incorporation of XML and other architectures into software applications, such as MS Word, and PDF, the creation of metadata by creators should become easier and more automatic.

Ingest: Acquisition and Collection Development

The first function to be performed by the archive itself is acquisition, or ingest. This is the stage at which the created object is "incorporated" physically or virtually into the archive. The acquisition of electronic information for archiving involves the development of collection policies and gathering procedures, and these policies and procedures should be considered in tandem with the development of archiving system requirements.

Data Management

Metadata is needed to preserve the object and for users in the future to find and access it. Metadata supports organization, preservation and long-term access. Archiving and preservation require special metadata elements to track the lineage of a digital object, to detail its physical characteristics, and to document its behavior in order to reproduce it on future technologies. The major preservation projects had its own set of metadata that it considered important for preservation. In 2001, the Research Libraries Group and OCLC reviewed the various sets of preservation metadata and concluded that there was sufficient similarity among the elements that a core set of metadata for preservation could be identified. In 2001–2002, the Preservation Metadata Working Group developed a draft set of over 20 elements and numerous sub-elements for metadata preservation in the framework of the OAIS Reference Model. In order to gain consensus on this set and to provide operational and implementation guidance, a follow on group,

PREMIS, the Preservation Metadata: Implementation Strategies working group was formed. The draft element set for preservation metadata and the results of the implementation survey were published in 2004– 2005. The plan is to provide the preservation metadata set for testing and prototype implementations before moving the results into a standards process.

Formats for Preservation

Without a through understanding of the internal details of the formats in which digital objects are encoded, the long term preservation of the objects is not feasible. Specific instances of formatted objects must also be interpretable so that the significant properties of those objects can be retrieved. Most electronic journals, reference books, or reports use TIFF image files, PDF, or HTML. TIFF is the most prevalent for those organizations that are involved with conversion of paper issues of journals. For purely electronic documents, Adobe's PDF (Portable Document Format) is the most prevalent format. PDF provides a replica of the Postscript format of the document, but relies upon proprietary encoding technologies. While PDF is increasingly accepted, concerns remain for long-term preservation and it may not be accepted as a legal depository format, because it is a proprietary format. Therefore, Adobe, the Association for Information and Image Management (AIIM) and several other organizations have developed a draft standard for archival PDF, called PDF-A. This provides a file specification for a minimal set of PDF features and functions that will continue to be migrated from one version of PDF to another. The draft is currently in the ISO process.

Digital Triage : Developing Guidelines for What can and should be Saved

There should be informed skepticism about the claims of organizations that say they will archive the Internet. The library and archival communities already know that not everything can and should be saved. What is key is selecting which digital resources to preserve and which ones not to preserve. Librarians and archivists must develop digital collection development and evaluation guidelines to assist in deciding what can be saved and what should be saved, and what can't be, on a case-by-case basis.

Rescue Operations : Ensure Vital Electronic Documents are Preserved

Digital copying will not necessarily ensure the preservation of a digital document. The fact is that digital information may be outputted to microfilm for preservation purposes and it may even be appropriate to print an electronic document on acid-free paper and handle it according to established archival practices. Librarians and archivists need to work with industry to develop simple and cost effective print-to-microfilm systems; this will enable archives to preserve documentary collections that are provided in proprietary formats such as word-processors in a cost-effective fashion to be effectively preserved.

Being Legal

A library may have the rights to access and use electronic materials, but the right to preserve the materials may not be the same thing. Restrictions on access placed by rights-holders or by licensing arrangements for particular resources needs to be addressed when questioning whether the information can be preserved. Librarians need to work on contractual issues. There must be a concerted effort on the part of all libraries to work together to get the best contractual arrangements possible and to be more aware of the contractual issues associated with the licensing of electronic resources.

Promoting the importance of Preservation:

Librarians and archivists must engage in a concerted effort to raise the profile of preservation. Librarians and archivists need to encourage critical thinking and be highly pragmatic about the nature of the new medium and the challenges of digital preservation. The problem is that "digitization may come to be regarded as a panacea for all of the real and imagined problems libraries now face in connection with the preservation of physical collections: the growing need for storage space, the deterioration of books due to acid paper, and the rising costs of library operation. Only by increasing public support and understanding of the issues of preservation, both digital and physical collections can we hope to address the shortfall in fiscal and human resources that continue to hinder preservation efforts and impact upon library services.

Working Together

Sharing Mechanism is very important for librarians and archivists in this reference. Few libraries will be positioned to effectively archive large quantities of electronic information. Any given library will necessarily be required to select resources that they can archive and preserve according to their particular mandates and user requirements. In many cases, it will not make sense to duplicate efforts. Digital preservation efforts will need to be coordinated. In other situations, it may make perfect sense to duplicate archival efforts, particularly if the information is too valuable for historic purposes to be entrusted to only one institution.

Digital Preservation as Public Good

Librarians and archivists protect the public interest by making information available to the community and by asserting the importance of maintaining a record of our collective intellectual heritage. This task will be a continuing challenge because libraries and archives are too often considered to be competitors to publishers, document delivery services, and other private sector content providers.

Conclusion

Archiving and related issues of digital preservation are becoming ever more significant within the scientific and scholarly communication chain. Recently, several approaches for digital preservation have been identified and presented. Conventional methods are mainly technology emulation, information migration, and encapsulation. However, there is a lack of proven preservation methods to ensure long term safe preservation for digital objects. The issue of the copyright of intellectual and intangible properties is also a problem towards digital preservation. There is also a burning question: what is appropriate material for preservation and what can be 'edited out'.

Thus the challenges to digital preservation are considerable and will require a concerted effort on the part of librarians and archivists to rise up to these challenges and assert in public forums the importance of protecting a fragile digital heritage.

References

1. Digital Library Federation. http://www.diglib.org/preserve.htm.
2. http://archive.ifla.org/IV/ifla63/63kuny1.pdf.
3. http://shodhganga.inflibnet.ac.in/dxml/bitstream/handle/1944/1258/265-274.pdf?sequence=1
4. https://www.rimp.gov.ab.ca/publications/pdf/DigitalPreservationResGuide.pdf
5. **Beagrie, N. and Jones, M.** (2001). "*Preservation Management of Digital Materials : A Handbook*". London: The British Library.
6. http://www.imaginar.org/dppd/DPPD/126%20pp%20Intellectual%20Preservation.pdf
7. http://ftp.rta.nato.int/public//PubFullText/RTO/EN/RTO-EN-026///EN-026-07.pdf
8. http://www.setecinvestigations.com/resources/whitepapers/Preservation_Orders.pdf
9. http://www.nascio.org/publications/documents/NASCIO-RecordsManagement.pdf
10. http://www.sherpa.ac.uk/documents/D4-4_Preservation_Selection_Criteria.pdf
11. http://www.jisc.ac.uk/media/documents/programmes/preservation/elopres.pdf
12. http://www.digitalpreservation.gov/library/docs/es_e-books.pdf

20

The Value of Print in Digital Era

Mrs. Kumkum Agrawal

Introduction

The main purpose of a library is to serve the needs not only of today's users but also tomorrow's. It follows that the library must have an enduring collection of resources that is accessible and meaningful to both current and future scholars.

Information communication technology changed the scenario of libraries all over the world. The library services are changed drastically. The availability of Internet gives access to information relating to almost every aspect of life, education, research and other spheres of knowledge. Powerful knowledge management, software have changed the way literature is to be organized, stored, accessed and retrieved. Web has made availability of information to all one who has the knowledge, skills and experience to apply this everchanging fast technology, can only survive in today's competitive environment. In fact, we are heading towards virtual libraries.

Print still dominate in scholarly publication. Physical formats will remain in collections for many years to come. As radio is replaced by television, the creation of a new medium does not necessarily invalidate the former ones. Print is a time-tested format that continues to fulfill promises that technology can not yet deliver.

A common fallacy not everything is on-line is that all information is available on the Internet, whether free or through a fee-based service. The majority of the world's published materials remain in physical formats only. Even in recent technology-friendly years, the printed books has not been made obsolete by e-publishing because most of the published titles were not available in electronic format. This shows that libraries must continue to examine, evaluate and collect print materials.

All of the Materials in Electronic Format are not Always Free

Not all online information is free to the public. Libraries subscribe to many electronic databases that individuals could never afford. With so much information available to Web surfers, it appears as if not cost is the norm for e-resources. However, libraries can and do arrange for seamless authentication to fee-based resources, thereby eliminating the requirement for individual user passwords and identification information; thus a resource appears 'free' to end users. Such users are often unaware of any restriction to the resource until they attempt direct remote access and are prompted for a password.

Licensing

Once purchased print materials belong to a library, on the other side many electronic products are licensed. A library has access to licensed contents only so long as it maintains and pays for a subscription. This distinction plays a significant role in the analysis of the long-term value of a resource to a library.

Authentication

Books and journals found in libraries will have been published under rigorous guidelines of citation and accuracy and are thereby allowed into libraries collection. These standards are simply not imposed on websites. They can show up in search results whether or not they provide citation. With enough research, the accuracy of web resources often can be determined. But it is very time consuming. Libraries make research much more efficient.

Copyright

Some authors and publishers not allow their work to be freely accessible over the Internet. It is already prohibited by law to make copyrighted books fully accessible through Google book search. Current copyright law protects works for 70 years beyond the death of the author.

Preservation

Print materials are also subject to mutilation but in such instances the original text itself is unchanged, only the copy that the individual library holds is damaged. With e-resources the source document can be modified as easily as a copy, without indication as to when the changed was made, who authorized it, and what motivated the alteration. Destruction of print materials results in the removal of an individual volume from collection, destruction of an e-resource corrupts the source document, making verification or retrieval of the document by later generations impossible.

Print text have life spans counted by centuries, determined by paper quality and the elements that touch it, and extended by the preservation techniques. E-documents' lives are briefer, due in part to the absence of set standards in digital production.

Web sites go offline or their addresses change other sites that point to these resources could easily and unwittingly house a number of broken links. These sites can remain unedited for years.

Libraries on the other hand, have a well-accounted for stock of availáble resources and a standard indexing system that will deliver stable, reliable results consistently.

Advantage

At last I will point out some advantages of paper books:—

(*a*) They are easily obtainable (Book stores are everywhere).

(*b*) They are easily portable.

(*c*) They don't normally cause significant eye-strain.

(*d*) They are cheap.

(*e*) Text books are mostly in print form.

(*f*) Paper books don't need power to function. They can be read anywhere with sufficient light.

Conclusion

This shows that news, journals, books and other resources can be accessed through the library on Internet. Technology is integrating itself into the library system, not replacing it. Libraries should adopt to social and technological changes, but they can't be replaced. While libraries are distinct from the Internet, Librarians are the most suited professionals to guide scholars towards a better understanding of how to find valuable information online. Indeed a lot of information is online. But a lot is still on paper. That is why most of the academic libraries are considered as "Hybrid Libraries" which are providing information both from e-resources and printed documents.

References

1. **A. Amudhavalli and Dr. Jasmer Singh Ed.** (2010) : "*Challenges and changes in Librarianship*", Vol.1, Ist ed.,B.R. Publishing Corporation, New Delhi
2. **A. Amudhavalli and Dr. Jasmer Singh Ed.** (2010) : "*Challenges and changes in Librarianship*", Vol. 2, Ist ed.,B.R. Publishing Corporation, New Delhi
3. **Michelle M. Wu**, "*Why Print and Electronic Resources are Essential to the Academic Law Library*".

21
University Library Collection Development Resources and Policies in an Electronic Age

Rajesh Kumar Singh

Abstract

Collection Development is a vital process in creating and building a library collection. This article discusses the definition of collection development in electronic age, on line collection development resources, collection development policy, selection criteria for offline e-resources, websites and web resources. Author also explores the format selection analysis, Functionality and Formats, Longevity and issues relating to cost are the new phenomena in this electronic era.

Keywords: *Electronic Resources, Collection Development Policy, Collection Development, Open Access Initiatives.*

Introduction

The new advances in information communication technology, networking, use of internet and electronic Products have brought about a radical change profoundly affecting the university library's landscape. It has affected the selection, acquisition and information transfer process. The Technology is mainly being used for communication, database searching, bibliographic and full text searching. It has also changed the concept of archiving. The organization of information, storage, access, preservation and retrieval has become both simplified as well as complicated. It is believed that information has become more garmented, piecemeal and disembodied, resulting into changing its face completely. According to (Harloe and Budd, 1994) "Economic forces and technological advances have combined together to create a new environment, where access to collective scholarly resources that no library could be ever afford, supersedes the historic quest for the great comprehensive collection" Swan (1992) has given the view, "we are no longer accessing the whole fabric of information, rather bits of data, sound bites and images torn from it". IT has provided several new media, new modes of studies, organizing retrieving and communicating information to users of the same. This ability to communicate has opened up the emergence of new modes of publication from authorship to digital system for information access which is easily movable and transferable to any point on the globe. The developments that are emerged during the present century have brought many opportunities and challenges for the libraries in general and library professionals in particular. The most important issues or realities are electronic publishing and networking of libraries, changing concept from ownership to access, and commercial availability of databases, etc. These realities have a direct impact on libraries. Hence, library professionals must cope up with these realities to meet the challenges. In the context of new developments in information technology mere collection of books is meaningless since information can be accessed through various networks. In the current electronic information environment emphasis is towards excellent collection development than large collection and developing effective means of getting access to remote databases.

Definitions of Collection Development

The collection management is a fundamental concern. (Demas, 1994) puts the matter into perspective this way:

"Electronic publishing has profound implications for collection development, which is defined as the intentional and systematic building of the set of information resources to which the library provides access. While the principles of collection development, which were developed in the world of print publications, do not change radically with new publishing technologies, methods of decision making and specific selection guidelines must be adjusted significantly to incorporate new publishing formats.

- A planned process of selecting and acquiring library materials to meet the needs of a library's community; cooperative collection development refers to a group of libraries working together.

  [www.mlin.lib.ma.us/mblc /public_advisory/trustee_handbook/go01.html]

- Term used to describe the activities involved in selecting library materials for purchase.

  [www.wpi.edu/Academics/Library/Help/glossary.html]

- The process of identifying the strengths and weaknesses of a library's information resources with respect to patron needs and community resources, with an aim at maintaining the strengths and correcting the weaknesses; the process requires a constant study of patron needs and changes in the community the library serves.

  [tsolutions.net/help/glossary.htm]

- Process of acquiring resources to build a collection that meets the needs of the curriculum and the instructional process.

  [mts.admin.wsfcs.k12.nc.us/admin/techplan/techpl16.html]

- The process of examining the academic and research needs, and selecting materials in support of those needs, for both faculty and students.

  [www.nyu.edu/library/bobst/research/tsd/glossary.htm]

Collection Development Resources

Some resources on University library collection development:

- Acqweb's Directory of Collection Development Policies on the Web - Directory of Collection Development Policies on the Web. This Directory of Collection is edited by Beth Mazin, Assistant Director, Memorial Hall Library, Andover, Massachusetts.

  [*http://acqweb.library.vanderbilt.edu/acqweb/cd_policy.html*]

- BUBL LINK / 5:15 Catalogue of Internet Resources: Library and Information Science. Located at Andersonian Library of the University of Strathclyde in Glasgow, Scotla nd. All items are evaluated, catalogued, and described, and all links are checked monthly.

  [http://bubl.ac.uk/link/lis.html]

- Dartmouth College Library Collection Management & Development Program : Guidelines for Writing Collection Development Policies.

  [www.dartmouth.edu/~cmdc/bibapp/cdpguide.html]

- Federal Depository Library Program: Collection Development Manual provides assistance to build the depository collection.

  [http://www.access.gpo.gov/su_docs/fdlp/coll-dev/]

- Federal Depository Library Manual Supplement: Collection Development Guidelines for Selective Federal Depository Libraries, September 1994.

  [http://www.access.gpo.gov/su_docs/fdlp/pubs/fdlm/coldev.html]

- GODORT Handout Exchange Collection Development Policies. American Library Association Government Documents Round Table Education Committee. Provides links to Collection Development Policies.

  [http://www.lib.umich.edu/govdocs/godort/collec.htm

- Internet Resources for Librarians. Created by Vianne Tang Sha. The site's goal is to "provide a one -stop shopping center for librarians to locate Internet resources related to their profession".

  [http://www.itcompany.com/inforetriever/index.htm]
- Library of Congress on "Collectio n Development and Internet". A Handbook A Brief Handbook for Recommending Officers in the Humanities and Social Sciences Division at the Library of Congress Compiled by Abby Yochelson, Cassy Ammen, John Guidas, Sheridan Harvey, Carolyn Larson, Margo McGinnis. (Last updated in June 1997)

  [http://www.loc.gov/acq/colldev/handbook.html]
- ODLIS: Online Dictionary of Library and Information Science: Authored by Joan Reitz is a comprehensive dictionary of Library and Information Science terms. Many collection development terms are included. This site is an excellent resource for the student of collection development and other areas of librarianship

  [http://www.wcsu.edu/library/odlis.html]

Policy of Collection Development

Before the electronic revolution, Richard Gardner (1981) published a book on library collections, in that book he has mentioned four basic criteria for selection viz. quality, library relevancy, aesthetic and technical aspects, and cost. But over time the meanings of some of these concepts have changed and the context in which they operate has also changed drastically.

Format selection is a complex process. Ultimately, selection decisions will vary depending on a library's specific needs as outlined by its mission statement and collection development policy. Thus, library professionals should determine some of these following options while selecting the formats.

Format Selection Analysis

In determining the optional format for a particular resource, librarians should first consult their organizations collection development policies for selection guidelines. A collection development policy is a fluid document, it must evolve as new formats are introduced as well as reflect the changing needs of its users. A Collection Development Policy provides general and specific format selection criteria (e.g. print, microforms, Internet resources, online databases, CD-ROMs, floppy disks, and e-books). A well written Collection Development Policy typically identifies subject areas in which it may be preferable to have one format over another (Lee and Wu, 2002).The opinion in The New York Times vs. Jonathan Tasini (United States Supreme Court, 2001) set forth the rule that publishers owning the rights to a print publication are not automatically entitled to the electronic rights to the same article. An online database may exclude information that is readily available in its equivalent print resource (Carlson, 2002).

Functionality and Formats

One of the primary reasons libraries subscribe to an electronic resources is functionality. Web databases can provide support in ways that print and microform cannot duplicate. Online resources are also generally easier to navigate; users can jump from one page to another or one section to other with a simple click of the mouse instead of underlying the more labour intensive method of searching through various print volumes to arrive at a section number (Hogan, 1998).

Longevity

An important and unavoidable factor in evaluating formats is determining longevity. Libraries must balance the needs of current users with those of future generations. They have the responsibility of preserving information so that articles will still be available after a publisher is out of business or an article is out of print. Longevity is judged by several factors like the ability of the product to withstand ordinary use within a library setting, the continuing availability of the article in the foreseeable future, and the needs of primary users.

The Essential Issue: Cost

Price is both simplest and trickiest factor. It is simple because, for most libraries, their budget will determine whether or not to go for purchase. However, with online products, the analysis is less straight forward. Not only do publishers have different pricing structures based on the type of license, but they can also have different pricing schemes based on the type or size of library. Publishers can offer discounts if a library purchases titles in multiple formats, or if it purchases a certain cluster/bundle of products. Although these types of deals are also offered with print products, especially volume discounts for large purchases, electronic pricing varies more widely than print pricing (Kevil, 1997). The policy document should be very clear and suggestive for the information professionals either to select or reject the print or non-print media.

Selection Criteria for Web Sites

Wyatt (1997), Spiller (2000) and Singh (2003), offer criteria for evaluating websites.

Credibility, conflicts of interest

- Web site owner or sponsor, conflicts of interest
- Web site author, credentials

Structure and content of web site

- References to sources
- Coverage, accuracy of information content
- Currency and Timeliness of content material
- Readability of material
- Quality of links to other sites
- Media used to communicate information.
- Functions of web site

- Accessibility of site via search engines
- Use of site, profile of users
- Navigation through material
- Archiving
- Arrangement or design of the title and its content.

Selection Criteria for Web Resources

The same criteria are applied to the selection process for electronic resources as are applied to print. In addition to standard selection criteria, the following criteria specific to electronic resources should be considered.

- Wider access and greater flexibility in searching
- Availability in a multiplicity of formats (e.g. ASCII, PDF, HTML, SGML, etc.).
- Electronic resources should be available before or not later than the publications of the article in its print format.
- User friendliness.
- Publisher/aggregators reliability and customer support.
- Increased functionality.
- Enhanced access to remote users.
- Time availability
- Hardware and software requirements should be taken into consideration.
- A trail period is potentially available for examining the utility and the value of the resource before a final commitment or order is made with the publisher/vendor.
- Service Implications .

Selection Criteria for Offline E-Resources

Chowdhury (1999), Mambretti (1998) and Rowley and Stack (1997) suggest evaluations based on:-

- The necessary amount of staff time to provide access, training and assistance.

- The long term viability of resources for preservation purposes.
- The long term usability of a resource's data (for a specific period of time).
- The broad accessibility of the resource under present copyright laws and licensing agreements.
- The compatibility of hardware.
- The availability and adequacy of documentation.
- The currency of the resource's information, if deemed necessary for subject matter.
- The user friendliness of the resource.
- Network Capability.
- The replacement policy of the publisher in the event of damage.

Conclusion

In electronic environment, university library collection development is becoming collection management, which is much wider in scope. The librarian must act as a knowledge manager, applying the skills right from collection planning, selection, analysis and cooperation in order to manage the intersection of both print and e-resources. They need to think about the availability and accessibility of multiple electronic formats in order to deliver the best information to all users in the least possible time. The web has introduced new resources to collection managers throughout the world. In the changed environment today, the duty of a librarian is to expand the range of resources for the benefit of users especially to include those available in electronic format, viz. web-based or web accessible information resources. The general pattern of professional activities remains recognizably similar to what it has been for the last half century and more. Librarians still evaluate information resources, connect users to the information they need, and organize information for easier access by the users. With the advent of web based resources, librarians we are finding that their role as information intermediaries demands a new sub-set of quasi technical skills and awareness. Librarians must not only identify and facilitate access to electronic information resources; but also educate library users about how to access them and when to use them Looking into the changes that are taking place in the developed countries like U.S.A., U.K. and Australia etc. in the filed of information selection, storage, they are providing access to different types of users in an electronic environment with effective manner. The developing countries like India are lagging behind because of the poor infrastructure facilities. In order to cope up with the situation Information Professionals should go together with the technology. Information technology has drastically changed the very nature of work and functions of librarians and made to depend more on hardware, software and networking etc. This requires lot of expertise, knowledge, skills including the training for the professionals to render effective and efficient services to their clientele in a changed environment.

References

1. **Carlson, S.** (2002). Once Trustworthy Newspaper Databases have become Unreliable and Frustrating : Supreme Court Decision Led Publisher to Purge much Archival Material, Chronicle of Higher Education. A 29.
2. **Chowdhury, G.G.** (1999). "*Introduction to Modern Information Retrieval*". London: Library Association.
3. **Demas, S.** (1994). "*Collection Development for the Electronic Library : A Conceptual and Organizational Model*". Library Hi-Tech, 12(3), 71-80.
4. **Gardner, R.K.** (1981). "*Library Collections : Their Origin, Selection and Development*". New York:McGraw Hill.
5. **Harloe, B. and Budd, J.M.** (1994). "*Collection Development and Scholarly Communication in the Era of Electronic Access*". Journal of Academic Librarianship. Vol. 20 (2), pp 83-87.
6. **Hogan, J.** (1998). "*Why won't My Westlaw search work on Lycos*"? Perspectives: Legal research and writing. Vol. 7 (3), pp 123-126.

7. **Kevil, L.H.** (1997). "*Payment and Subscription Models for Online Publications*". Library Acquisitions:Practice and Theory. Vol. 21 (3), pp 247-249.

8. **Lee, Leslie A. and Wu, Michelle M.** (2002). "*Do Librarians Dream of Electronic Serials*"? A Beginner's Guide to Format Selection. The Bottom Line: Managing Finances. Vol. 15 (3), pp 102-109.

9. **Rowley, J. and Stack, F.** (1997). "*The Evaluation of Interface Design on CD-ROMs*". Online and CDROM Review. Vol. 21 (1), pp 3-13.

10. **Singh, S.P.** (2003). "*Evaluation of Electronic Reference Sources*". DESIDOC Bulletin of InformationTechnology. Vol. 23 (2), pp 43-47.

11. **Spiller, D.** (2000). "*Providing Materials for Library Users*". London: Library Association Publishing.Strategic and practical considerations for signing electronic information delivery agreements.Available at: http://www.arl.org/scomm/licensing/licbooklet.html Accessed Date: (31/09/2011)

12. **Swan, J.** (1992). "*Rehumanising Information : An Alternative Future*". In J. Hannigan, Ed. Library (J) Literature 21: The best of 1990. London: Scarecrow. pp 18-29.

13. **Wyatt, Jeremy C.** (1997). "*Commentary : Measuring Quality and Impact of the World Wide Web*". BMJ.Vol. 314 (7078), Website: http://bmjjournals.com/archive/7078ip2.htm (Access Date: 31/03/2011).

22

Print and E-Resources : A Comparative Discussion with Special Reference to Legal Resources

Mrs. Shiva Parihar

Abstract

Libraries are the knowledge gateways for the users and information brokers. It is necessary for the administration first to check the convenience for the users then the space allocation and budgeting while selecting the resource for library. To decide this there are some necessary factors to be checked while electing between the e-resource or the print. For law libraries it is important to select the resources with special care because neither print nor electronic resource can be ignored both are important for the successful survival of law libraries.This paper is an attempt to find out few criteria under which law libraries can compare the resource and select accordingly.

Keywords : *E-resource, Print Resource, Searching, Authenticity, Preservation, Reliability, Ease of use.*

Introduction

New technology has brought a new era of electronic resources to the world of knowledge. Convenience, cost, timeliness are the things that have entered to the knowledge gateway of the library. The purpose of the library is to fulfill the needs of the users current as well as the future. It is necessary to ensure the availability of the resources whether e-electronic or print for a long time with authenticity, ease of use, technological stability and preservation. In law library now there are lot of e-resources to help the users but still it is necessary to check the reliability of the resource searching power and stability then only we can replace the print resources with the electronic one. Here there is a comparative discussion amongst some necessary points to be checked while selecting resources in between the print and electronic.

Searching

Technology has helped a lot the users to get the services in electronic format very easily. The e-resources has made easy the searching through different kinds of searching like Boolean searching, nested queries, limiting executed searches and direct linking to full text. As if in any case we know only the name of judges in the bench or the name of appellant or the respondent it is an easy task to get the whole referred case through legal databases like SCC online. On legal databases like Manupatra, SCC online, Lexis Nexis etc we can get the cases on different topics for last many years at one place from supreme courts or high courts in a very short span of time and in case of printed documents it will take time lot of effort to locate the cases from the respective year's volume or to go through the yearly digest like Supreme Court Yearly Digest or AIR manual. The search result comes with the highlighted term this makes easy to evaluate how much the result is relevant to us. All this may be useful but in the absence of exact citation out of many resultant cases that we got on the bases of some topic search first creates a kind of confusion and needs to be studied very carefully.

Ease of Use

There are many points which support e documents under the ease of use one of them is portability i.e.; the books occupies space, e-data also needs memory this is quite lesser than required by the books. The books together can't move together in number but e documents can be moved in USB or CDs. But still there is a problem of memory capacity and data loss due to some technical mishappenings to the data stored in some device.

Likewise another point is Malleability it can be messaged and used through cut paste method but this has also created the problem of "plagiarism".

Relational mobility similarly a characteristic which we can get in a better way in e-documents as the referred item in a printed document must be located on physical shelves which takes time .But in case of e-documents the user can get the cited documents through a click on hyperlinks for e.g., the cases cited in legal databases can be achieved via a click on the link given. This also depends on the users 'ability to search the references.

There are some other points to be discussed while we are taking ease of use in electronic documents in comparison to the printed one. For visually disabled patrons the legal databases like Manupatra and Westlaw the option to listen the case note is present without the help human interpreter or reader. But still this facility is not available in every database. Out of these facilities in e-resources we can notice some problems while using the legal databases like the background colours are sometimes not soothing so that it is none convenient for the users to read the matter on the screen for a long time for e.g. Manupatra is a database in which with the background colours and the format given is difficult to stay for a long time, whereas the format of SCC On line and Westlaw is really very convenient.

One of the most important characteristic of e resources is that number of unlimited users can get the access through IP access, while in print only one can use the particular resource. At the time of moots, seminars, presentation and projects this becomes very helpful.

Ownership

Once the library purchases the books or start the subscription of print journals the library owned the book for a very long time so the users can have the permanent access to that book. But in case of e-resources like databases of the online journals the access we got through license is temporary access only. The library has to pay a lot of charges to start the subscription and also at the time of renewal library has to pay for the old as well as the new added information. And if in any case institution fails to pay for the renewal of license immediately after the fixed period user will not get the access to the new but also will loss the old one. For example in case of legal databases once library fails to renew the subscription immediately the user lose the right to use the service i.e. the new added cases as well as the old one. Similarly incase of e-journals like "Economic and Political Weekly" from this year. Earlier they have the policy to provide their online achieves from the year 1966 to the Universities, those who are subscribing regularly their print journals but from year 2010 they have changed their policy and now they will provide this free of cost services to the colleges only comes under the 12(b) Clause of UGC. With this the library those were availing the facility have lost the whole achieves and for this again the users depending on the print volumes available in the library.

Once we get the ownership of the print material it is very easy to use it for inter library loan purpose under the established copyright laws and practices. In case of electronic products the license to use comes with the contract terms and conditions different from the print sources. Even the publisher is same but this does not work under the established copyright laws and practices. Most of the online databases fails to fulfill the inter library loan services amongst the library. After the access of license contents there are certain restrictions in case of e-resources as there are some downloads limits. But in case of print materials we can take the photocopy of the whole required materials.

Authenticity

Academic library needs to provide authentic data to its users especially in case of law libraries to support arguments and to improve scholarship. We can get a lot on any topic through internet provided by a different search engines like Google, Yahoo, MSN search etc. But there is no guarantee of these postings as there is no authority. Only some reputed domains like gov or .edu or .ac.in are having some authenticity. In comparison to these the printed resources regularly got edited and revised these changes are authentic due to known authorship.

E-resources Data that are fee based like legal databases Manupatra, SCC Online, LexisNexis etc. are reliable due to regular revision.

From legal perspective the Supreme Court of India has given the rule for citing the cases during arguments. Under this the court rely on the printed Supreme Court Cases for Supreme Court's decision and for High Courts and other lower courts All India Reports because these are very old and authentic. Opposite to that courts do not rely on Manupatra usually only if the citation is not available in any other authentic resource then the court goes for Manupatra. Other databases like Indlaw, Legal Pundit are not accepted in the court. But in case of government publications like law commission report published by Law Commission of India sign by authority and edited as per the changes would be accepted in the court and other places where authenticity is required.

Changing Standard & Preservation

The technology changes continuously due to change in technical standards and formats, with this the old becomes unusable to keep up to date with this situation and to keep digital resources accessible for use in future investment in continuation is required. A very practical example of technological changes in MS Word we can't get word 2000 in MS Word 2007 or MS 1998 in other standards. Access the e-resources we depends on equipments and technology. There are lot of technologies for video players like VLC, Real player, Classic Player etc. to have an access to the video similarly there are lot of changes in markup language from HTML to DHTML to XML use to prepare documents for web publication. For information format we have PDF, JPEG, TIFF etc. so far a library the resources should be selected on the bases of factors in future cost for migration for preservation. The life span of print text is of hundreds of years depends on the paper quality it is difficult to preserve the paper for government and court due to absence of set standards in digital production.

The printed word once printed is fixed and unchangeable. The content of printed is stable in preserving format. In comparison to this e-documents may or may not be accessible in future due to continuous change in technology and lack of active back up.

Conclusion

Through this comparative discussion we can say that whether the resource is printed or electronic the users should be benefited. The library should provide to users "The right tool at right time at right place to the right users. In law library it is necessary to check the authenticity of the resource provided because these would be the base for the arguments and judgments.

References

1. **Jain, H.C.**, "*Using Law Library, Legal Research and Methodology*", ILI, 241-263pp., 2006.
2. **Solan, Amy E.**, "*Basic Legal Research : Tools and Strategies*", Aspean, 3rd ed.(2006).
3. Supreme Court of India Practice and Procedure : A Handbook of Information, Supreme Court of India, 3rd ed. (2010).
4. **Waldman, M.**, "*Freshermen's Use of Library Electronic Resources and Self Efficacy*", at http://www.informationr.net/ir/8-2/paper150html last visited March 28,2011.
5. **Wu, Michelle M.**, "*Why Print and E Resources are Essential for Academic Law Libraries*", Law Library Journal,vol97:2,2005,234-256pp.

23

E-Resources : The Emerging Issues at a Glance

Punit Kumar Singh

Abstract

The last three decades are the era of the information explosion and information revolution. This information has been stored in the form of the print format and digital format. The print information has been changed in digital format by digitization. Thus the both digitized and digital born information constitutes the electronic resources. It has become a significant problem today with technological advancements and creating and storage of our work on digital formats. It is a challenge to imagine not only how to technically preserve e-resources but also how to choose what to be preserved and how to guarantee the electronic data reliability and authenticity in the future. The combined problems of immense volume, unstable storage media and obsolete hardware and software add up to some very tough problems, which have to be dealt with. This paper discussed with emerging issues regarding the handling, care and preservation of e-resources.

Keywords : *File formats, electronic record, e-resources, electronic record management, Archive management.*

Introduction

All libraries today create electronic files. Some have digitization programs, where they cover print or other analog materials to digital form. Most have an electronic catalog, of OPAC, of their collection. Many have created databases of internal or public use. All have prepared word processing document, spreadsheets, and other digital files. Even if we exclude formal digitization programs, all of us are electronic publishers in one way or another. At one time, fire, water and pests are great threats to a library's collection and records. Now they have been joined by other, more insidious, but just as disastrous threats: computer viruses, hackers, file format obsolescence, storage media degradation or obsolescence, platform dependence, catastrophic system failure, natural disasters, terrorist attacks, and simple neglect. Protecting our digital resources requires vigilance, education and planning.

Policy Issues

"The introduction of electronic journals has transformed scholarly communication in extraordinary ways–making it possible to disseminate research results more quickly, to provide hyperlinked access to cited publications, and to amplify text with images, audio and video files, datasets and software…" (Cantara, 2003). It requires pin pointed guidelines and policies to help tame the significant problem of preserving digital information. "Part of the challenge with digital preservation is recognizing the full breadth of the tasks facing information managers" (Hunter, 2002). The problem can not be simply solved. The choices are broad and must be considered carefully. How we act in preserving our data now will affect what information makes it into the future. There need to be procedures put into place that will give guidelines on how digital preservation can most effectively and cautiously be done. It is an overwhelming task and a great deal of thought, expertise, and experience must go into these guidelines; but the problem cannot be ignored.

The policies should be designed with the finished result in mind. Finally, the right questions need to be asked to determine how the policies should be developed. These questions include how the preserved digital documents will be used and what the best ways to go about ensuring the effectiveness of the digital preservation project. Technology advancement is moving swiftly and if information is to be retained, we must act now. We must set up policies that will preserve our heritage. The media we preserve our data on is just as important as the data itself. It will be instructive for future generations to see just how we have progressed in the different technologies we have chosen. Why we chose VHS over Beta and Beta over HTML. There will always be controversy about which information should be saved. It always costs money to preserve content and materials. A certain amount of weeding needs to be done. It is not necessary to save everything, but the weeding needs to be done with a light hand. What is unpopular or seems unimportant today may become very important in the future.

The sciences do not always know what will be useful information to the future, just as the humanities do not always appreciate a great work of art or excellent piece of music when it is first created. Policies of digital preservation should reflect this need for wide-reaching preservation. The policies for preserving the digital information also need to take into consideration how people will be able to access the information.

Storage Media Issues

Storage media and networks/infrastructure are vulnerable to instability, obsolescence, natural or man-made disasters, security breaches, viruses, hacking, improper environment, malfunction, improper maintenance, and human error. Generally speaking, you get what you pay for, so buy the very best you can afford whether you're purchasing CDs or security software. Hire competent people to manage and maintain your network. If they are knowledgeable and thorough, they will be able to avoid problems and will save you far more money than you pay them.

Types of storage media include hard disks, tape, and optical media. Hard disks are convenient and high capacity, but will last only about five years. Gold CDs are the best choice among optical media, and have a shelf life of five to ten years. DVDs have a higher capacity (sometimes it is only possible to get one TIFF file on a CD) but have not yet been proved dependable for long-term storage. Never trust your backups to cheap CDs or DVDs. How you store the media is as important as the media you choose. HDD are used now days. There will always be changes in information technology, and we must plan for it. Market forces encourage change, and today's hot technology is tomorrow's useless doorstop. The problem is compounded by the lack of standards for storage media and by the fact that there is no equivalent to the "open access" or non-proprietary software. We are at the mercy of the hardware manufacturers.

There has been a significant improvement in recent years, with a movement away from digital storage media, which are more fragile and less stable over time, towards the introduction of more stable and reliable media. The improved options for long-term storage of digital information include areas such as ion-milling and holographic media.

The International Council on Archives (ICA) Guide to Managing Electronic Records sets out seven criteria for media used for preserving electronic records (ICA, 1997):

1. Open standards for digital recording on the medium;
2. Robust methods for preventing, detecting and reporting errors;
3. Sufficient market penetration;
4. Known longevity;
5. Known susceptibility to degradation or deterioration;
6. A favorable cost/benefit ratio; and
7. Availability of methods for recovering from loss.

The current best practices in digital preservation employed by many of the archival institutions include:

- Selecting storage media most appropriate for long-term data retention;
- Converting data to standard formats to facilitate its processing on a variety of computing platforms;
- Migrating data to new technology platforms when the computing environment is upgraded;
- Preserving systems documentation required to process the data copying or recopying the data onto new storage media at regular intervals; and
- Taking steps to store and maintain these media properly.

These measures are labor-intensive, error-prone and costly and must be performed periodically for as long as the data are retained, making the long-term preservation of digital data a substantial undertaking.

Media Decay Issues

One solution to the problem of media decay is to copy the data to newer storage media but this kind of migration is not perfect. Migration of data to the latest media and software versions can be done on a two to three-year cycle but will require a significant monetary investment for each conversion, constant human attention and personnel training.

Digital migration requires the staff to have knowledge of the old formats as well as the new, the ability to analyze and recommend the best of new formats, the time to implement and test migration pilot programs, and the capacity to develop and continually refine migration processes. Moreover, during each conversion, the file is susceptible to corruption. Formatting can shift and data can drop out bit by bit over time. This can result from machine, software or human error leading to strange renderings of documents from which the original information cannot be recovered. Ideal policies, standards and strategies need to be formulated to ensure that authentic electronic records can be preserved over long periods of time.

Technological Issues

"No single method has been identified to preserve access to physical format digital material. It will probably require a combination of different methods..." (PADI). There is "no single solution" where one select digital preservation technique will work best for preserving all digital documents (Kranch, 1998). There needs to be a variety of possibilities that can be used dependent on the situation at hand. One possibility of preserving digital information is to print out the information on paper and store it that way (Hunter, 2002).This approach can not be applicable because it cannot be possible to store all information on paper for example the hyperlinked information, audio-visual files etc.

Another preservation technique is to set up standards on how every digital item should be preserved. One example of this procedure would be to use an ASCII format to store the digital information (Hunter, 2002). This standardization process has some downfalls to it though. ASCII could not be used to preserve all language characters. The digital preservation process needs to include information from all languages. Other standardization processes have similar problems.

Another technique in preserving digital data is to preserve all the software and hardware that are created so that old information can be read on these old computers (Hunter, 2002). The benefit to this approach is that future generations would have the opportunity to see first hand how the technology worked, not just see the machines and then read about how they functioned. The problem with this approach is how hard it would be to keep them working over time. Preserving digital information can also be done on a new computer that is designed to function like an older one (PADI, n.d.).

"Digital materials are surprisingly fragile. They depend for their continued viability on technologies that undergo rapid and continual change" (Flecker, 2000). Because of this fragility and diversity in product, a variety of preservation techniques must be used in storing our digital data.

All of the techniques for preserving digital data that have been discussed have their benefits and their problems. It is important to find which preservation technique is going to work best for the individual information item and then use it. Varied preservation practices need to be used so the digital information gets the most effective form of preservation. We need a variety of preservation techniques to process the digital information. One method of preserving all digital data is not going to work. Each method has its own problems and none of the solutions will work well for every item being processed every time. There have to be many techniques in our preservation processes. Without them it will not be done properly.

Approaches to Preserving Electronic Records

Current initiatives are pursuing a variety of approaches that can be put broadly into five categories:

1. Preserving the original technology used to create or store the records;
2. Emulating the original technology on new platforms;
3. Migrating the software necessary to retrieve, deliver, and use the records;
4. Migrating the records to up-to-date formats; and
5. Converting records to standard forms.

Each of these methods has pros and cons and none of them is entirely satisfactory.

Preserving the Original Technology

Stephens (2000) suggested the adoption of the museum approach to preserving digital data. Such an approach, according to Stephens, requires the fulfillment of the following:

1. Digital archives should be transcribed every 10 to 20 years;
2. Recording systems and media should be preserved; as should;
3. System hardware and software;
4. Operating system;

5. Operation manuals; and
6. Ample spare parts.

However, saving these processing components indefinitely may not be necessary.

Advocacy

We have the foundation of what those have learned before us and we add to that wide base of knowledge with our own experiences, expertise, and perspectives. The Einstein and Shakespeares of today may be using some form of digital formats to save their great works, and we want to be able to save them for future generations. To do this we need advocacy.

Advocacy is an important step in saving our digital preservation heritage. The public will not realize the importance of saving this valuable information over time unless they are educated on the impact of losing it.

Advocacy can be promoted by the formulation of the partnerships and networking among the groups that have a vested interest in preserving digital information.

The problems with preserving information are that we do not know how to best keep up with technology; and economics are a big factor in determining what our priorities for preserving digitalized information will be. New hardware and software are being created all the time. Advancement in technology is mind boggling. We are so excited by what is new or coming up on the horizon, that we do not take a step back to make sure the information stored digitally today is being preserved for the future. Advocacy becomes an important element in the solution to these problems because if the public is made aware of the possible loss, it will want to act to help save our digital cultural heritage.

File Formats Issues

File formats may be proprietary or non-proprietary. A proprietary file format is one developed by specific company and protected under intellectual property laws, whether patent or copyright. Most of the files we use and create on daily basis are in a proprietary format, whether it's MS Word, WordPerfect, or PDF. There is one critical difference among the proprietary files of different companies: some, such as Microsoft files, have a closed specification, which means that the company does not share the code that would allow someone to read the file without using their software. You need PowerPoint software to open a PowerPoint presentation. Other proprietary files, such as Adobe's PDF and TIFF formats, have open specifications, which mean that the company makes their code available to software developers and to their users. This becomes important if a company goes out of business, discontinues a product line, or stops supporting older versions of a file format. If the specification is open, it is still possible to get into the file, extract the information it contains and move it into a new file format. If a specification is closed, this "orphan" file is no longer readable or usable, and the information it contains is lost. Those of a certain age will recall WordStar, a popular word processing program in the early days of personal computing. When the company went out of business, there was no longer any support for those files, and soon there was so software available that would open them.

Standard file formats are certified by the ISO or another standard-making body. These files are platform and software independent; they do not require a certain brand of software or a particular operating system to be usable. ASCII text, XML, and SGML are examples of standard file formats. Work is underway to develop a standard for PDF files. When choosing a file format, the safest choice for preservation purposes is a standard. While there are no absolute guarantees, standards are the best insurance we have, since bits and bytes can degrade over time. If you must use a proprietary format, prefer one with open specifications. If they are widely implemented, like PDF and TIFF (which is often treated as a standard for image files), they become a de facto standard, and you can be relatively certain you will be able to read them five years from now. For any proprietary file, make sure that your file does not fall more than two versions behind the current software. Software manufacturers do not always make new versions backward compatible with files created with older versions of the software for the simple economic reason that they want users to upgrade. If you have word processing files you want to keep but don't want to update when new versions are released, consider saving them as ASCII text. You will lose the formatting but you'll still have your content. File compression is another danger for preservation, because when files are compressed some information, bits and bytes, are lost. This is a frequent problem with image files like GIF and JPEG. Master image files should always be stored in a lossless format such as TIFF.

To-do list for file formats

1. Inventory all files and make a list of formats.
2. Migration of older materials to newer versions (Word processing files, spreadsheets, etc.).
3. Try to limit the number of file formats you support.
4. Watch the technology market for news affecting file formats (M&A, new standards, product updates)

Preservation of E-Resources

An electronic resource is preserved if and only if it continues to exist in a form that allows it to be retrieved, providing reliable and authentic evidence of the activity that produced the record. Demonstrating the authenticity of electronic records depends on verifying that:

1. The right data was put into storage properly;
2. Either nothing happened in storage to change this data or alternatively any changes in the data over time are insignificant;
3. All the right data and only the right data were retrieved from storage;
4. The retrieved data were subjected to an appropriate process; and
5. The processing was executed correctly to output an authentic reproduction of the record.

Conclusion

Stephens (2000) suggests that archivists, record managers, and other information management specialists need to reinvent their professional practices to ensure permanent or long-term preservation of electronic records.

Library staff should take utmost care to update the library web sites/blogs regularly. The periodic updating of web sites would ensure the provision of most current and up to date e-information. Libraries should take active participation in various e-journal consortiums for maximizing procurement of e-journals at minimal cost on sharing basis by which institutes can afford to balanced e-subscriptions without much financial burden. The digital world provides great opportunities as well as great risks for library professionals. Plenty of evidence exists of what appears to be an international concern that automated systems are out of control and need to be better managed. Library professionals are increasingly viewed as part of the solution, even if their role is not yet truly defined or clearly understood.

References

1. **Cantara, L.** (2003), "*Archiving Electronic Journals*", The Digital Library Federation and Council on Library and Information Resources", Washington, DC, www.diglib.org/preserve/ejp.htm, .
2. **Flecker, D.** (2000), "*Preserving Digital Periodicals*", Building A National Strategy for Preservation: Issues in Digital Media, Council on Library and Information Resources and Library of Congress, www.clir.org/pubs/reports/pub106/contents.html.
3. **Hunter, G.S.** (2002), "*The Digital Future : A Look Ahead*", Information Management Journal, Vol. 36 pp.1-70.
4. **ICA** (1997), "*Guide for Managing Electronic Records from Archival Perspective*", International Council on Archives, Paris.
5. **Kranch, D.A.** (1998), "*Beyond Migration : Preserving Electronic Documents with Digital Tablets*", Information Technology and Libraries, Vol. 17.

6. **National Mission for Manuscripts.** available at: http://www.namami.ac.in

   See also: (http://www.ignca.nic.in/nl002502.htm). (Accessed on: 11/01/2011)

7. **Preserving Access to Digital Information (PADI, n.d.)** "*Physical Format Digital Material*", National Library of Australia, Canberra, www.nla.gov.au/padi/topics/52.html.

8. **Stephens, D.O.** (2000), "*Digital Preservation : A Global Information Management Problem*", Information Management Journal, Vol. 34 No.3, pp.68-73.

24

E-Age Library Professionals

Vikas Madan &
Vijetacharya Malviya

Abstract

The role of library professionals in E-age library environment is changing throughout the world. The new technologies have facilitated the transformation of data into digital format. As the charges occurred in library environment, the library and information science professionals should be acquaintance with different skills. This paper tries to reflect role of library and information professionals in E-age or e-resource environment. The role of library professionals is charged and they have to act as E-library Professionals.

***Keywords :** E-information literacy, skill enhancement programme E-resources, Digital Libraries, Professional Competency of Librarian, Edgeless libraries.*

Introduction

Rapid advances in information processing, storage and communication technologies have revolutionised the role of world wide libraries in disseminating information services to their users. The library managers are required to cope up with the increasing demands of faculty, students and research scholars against all odds like dwindling budgets and decreasing staff. An organizing body is necessary to be formed with the librarian from each library and among them one will act as a chairperson on rotation bases to lead the total consortium. The early recognition of the librarian as a caretaker or custodian of books. Now a days librarians equipped themselves with the skill of handling the new technologies. Libraries and library professionals as a discipline also therefore undergoing change therefore information literacy is necessary.

Objectives

The main objective of this study is to know about the dependency of the teachers and research scholars on resources. Some of the major objectives as are follows:

(*a*) Study the purpose and frequency of using the electronic resources and services available in the library.

(*b*) Know the productivity and quality of information retrieved through e-resources.

(*c*) Know the role of library professionals in context of e-resources.

(*d*) Study the impact of e-resources and services on the users and libraries.

(*e*) To find out the main advantages of using e-resources.

(*f*) To identify the problems faced by the respondents while accessing the e-resources.

Type of E-Resources

- E-Thesis
- E-Books
- E-Maps
- E-Journals
- E-Newspaper
- E-Mail
- E-Manuscripts
- E-Bibliographic Databases

Advantages of E-Resources

(*a*) One can access it round the clock across geographical barriers. The e-resources get published or reach the subscribed much before their print counter part, it also reaches all its subscribers simultaneously.

(*b*) Another important advantages of E-resources is that more than one person can access it at a time. Article can be download and printed simultaneously by more than one reader, depending on access rights and permission. Multiple and remote access makes it available at one's desk this is a boon for a huge campus where there are hundreds of readers with many departments.

(*c*) Electronic resources offset the missing issue problem, if a particular volume of the print version of the resource is not complete, library staff can download and print all the articles available online or can share it in digital form till the time hard copy is supplied by publishers. The information about free access against the print subscription is usually available on their website and sometimes even in the publishers catalogue. Some publishers provide free online access to under developed countries and online access with nominal fee for developing countries.

(*d*) Access to electronic resources is very useful to both libraries and users as problems of missing issues or delay in receipt of issues can be over come.

Changing Role of Librarians

Change is the law of nature, the professional or the institution, which does not alter with the passage of time, always lags behind and goes into oblivion after some time the modern society is information society. The progress and development of any society is directly dependent upon the extent of adoption to a changing environment. The modern world is of information technology where we cannot find any area untouched with is effect. The library and information science field is no exception to this. Today library professionals are known as information personnel. This changing emphasis necessitates the information professionals to be exposed to the new technologies.

In the present technological/internet era the professionals have change themselves as the information profession is being changed. The role of the librarians is E.age environment is to help the users to find the information they require then provide them with the tools to access and use the E-resources for their individual needs Croth (1996). E-age librarians and information professionals should be able to manage the digital information system as this encompass the overall competencies necessary to create, store, analyze, organize retrieve and disseminate digital information in e-libraries.

Librarians have traditionally been concerned with certain functions in the print era. i.e. collection development and acquisition, classification and cataloguing, circulation, reference service, preservation, conservation and archiving. There is no denying that this is a global library environment in which librarians and information professionals are still finding their way. The role of librarian in the E-age is also exponentially growing as the internet and world wide web is changing. The rise of digitalized information is an opportunity to evaluate the role of librarians and leads to the emergence of a new breed of librarian.

The e-age librarian is a technology application leader who works with the other members of the information management team to design and evaluate systems for information access that meet user needs. E-age librarians are supposed to be the essential by responding with a sense of urgency to critical information needs, they provide the information edge for the knowledge based organization. E-age librarians require two types of competencies - professional competencies related to the special librarians knowledge in the areas of information resources, information access, technology management and research and the ability to use these areas of knowledge as a basis for providing library and information services and second one is personal competencies which represent a set of skills, attitudes and values that enables librarians to work efficiently, be good communicators focus on continuing leaning throughout their carriers and survive in the new world of work. In this way, they have work as library managers. In addition to being library manager they also have to act as Technical Processors and so on by taking care of information quality.

E-information Literacy

It is well recognized that the skill level of e-information literacy has a great deal to do with the final success of an e-information service of a institution. The general users of the libraries do not have adequate knowledge on e-resources and how to use them effectively. However the knowledge required about not just to technology, but the domain of the application and the skills needed to determine "what they need" and "how they use". The shift from purchasing important resources according to user requirements to one of being an effective service provider in promoting what we have subscribed. This is purely based on the importance of various skills which leads to the success of both the user activities and the service.

Need of E-information Literacy

E-information literacy has become the essential component for library and information science profession because in this era of world wide web (www) library users have to use the latest information sources and apply techniques of information retrieval having the skills of inherent interoperability. The e-information literacy has thus influenced the professional habitat of the entire LIS profession, the main reasons of e-information literacy for the present society may be stated as follows:

(*a*) Libraries expansion beyond walls.

(*b*) Emergence of data mining and absolute web based system.

(*c*) Faster delivery of information.

(*d*) Interdisciplinary growth of literatures.

(*e*) Increasing of quantity and variety of formats of recorded information.

(*f*) Scenario of information and communication technology.

Having entered electronic environments the expectation from upcoming librarians is quite high and complex. They are expected to have requisite level and depth of information technology knowledge and skills in the modern E-world. Information technology related skill required by them in the e-information concern means different levels of literacy. Skill required for handling IT products, such as computer operating software, telecommunication products, data file management, DTP, world processing etc. The next level of e-information literacy include skill requires to apply information technology for service management and information processing search and retrieval. This involves collection and organization of data in electronic form indexing techniques searching as well as CD-ROD databases. In E-age librarianship requires a number of skills, tools and techniques, which are also required to manage traditional librarianship.

A well thought strategies could be worked out to promote e-information literacy through various programs at various level.

(*a*) Government policy for national information literacy for educating the use and digital information formats.

(*b*) Combining the information literacy mission with the curriculum at schools.

(*c*) Creation of dedicated website with interlinking provision facilitating the clearness.

(*d*) Organizing forms for campaigning about the needs of information literacy in the changing scenario.

(*e*) Professional library associations/ societies to play vital role launching the information literacy movement at national, regional and local levels.

(*f*) Guidelines and norms shall be decided by the State and Central Government the control of right to information act.

E-information literacy has become a global phenomenon, efforts are being made world wide to make people e-information literate and make them gain skills to their preparedness for life long learning and effectively utilize the super slow of digital information.

Updating the Skills of Library Staff

The basis goal of library and information profession has always been to provide access to information to those who need it. The activities realizing this goal have evolved and transformed over the years this includes. Available technology and need of an evolving information society. To manage the changing library environment, the library staff will need to be trained in the application of necessary tools in their work environment. Information activities have been guided by the developments in the field of storages, presentation and archiving of knowledge, collection development and organization of knowledge, information explosion and computers in information retrieval. Librarian and information professional involved in information gathering, storage, retrieval and dissemination on one hand and on the other hand the computer specialists who supports the library and informational professionals in this endeavour.

The modern storage media like, CD-ROD, microfilms, micro text optical laser, disc floppy discs, magnetic tapes and discs etc., play a vital role in the ever growing information world daily routines and functions of the library have been impacted and influenced very much by the modern communication, telecommunication, online network communication etc. So it is very essential to provide approximate training to the library personal. To handle and utilize the modern facilities librarian must have the knowledge and skills about the technologies and communication channels and should provide proper training to the staff for successful implementation of digital library, it is essential that LIS professionals are well trained and possess requisite knowledge and skills in this respect.

1. Librarians need to know understand - Knowledge resources (books, journals internet):
   - Teleological facilities and resources (computer, online catalogues, websites, LANs file servers etc.).
   - Financial resources, human resources, budget.
2. Competencies that required possessing in LIS professionals:
   - Acceptance of change.
   - Knowledge of user interaction with knowledge resources.
   - Provide quality service.
   - Possess excellent communication skills, constantly update personal knowledge base by keeping in touch with the latest development.
   - Create awareness among the users, make them accept the changes.
3. Technical knowledge required:
   - Operating systems - Windows, UNIX, LINUX
   - World Processing, Graphics, Spread Sheet and Presentations.
   - Database management system including the skills in bibliographic database management systems.
   - General purpose programming networking.
   - Web page development and content management.
   - Library software packages, acquaintances with digital libraries tools.
   - Seminar, conferences and workshops is being organize by the said institution.

Present Scenario in India

Besides the department of library and information science of different universities and institutions some other organizations have also been working for the skill enhancement of library professionals in India.

INFLIBNET also regularly organizes soul training programme and ILMS training programme for the working library professional, for making them technologically skilled so that they can use the soul package in their respective libraries in a proper way and face no problems. Planner is another important activity of INFLIBNET to make the library professional of India. Conscious about the advance application of technology in the library automation and networking.

From last few years, Central Reference Library, Kolkata has been organizing an apprenticeship programmes for the library and information science students. In this one month training programme a group of LIS students, each batch generally containing five students representing from different universities of India, are practically trained to work in a real library environment and to use software packages i.e. LIBSXS, GIST etc. This type of "earn and learn" training is very beneficial both for the students and the library and information science profession itself.

Indian association for special libraries and information centers is also working for improving the technical efficiency of the library and information science professionals. NISCAIR organizes different IT related short term courses for enhancing skills among the professionals. NISSAT encourages and supports variety of skill development programmes for the library and information professionals on CDS, ISIS, WINISIS, TQM, Internet and eb design etc.

DESIDOC and NASSDOC, two national documentation centers also arrange different short term training programmes for the professionals to acquaint them with the latest advancements in information and communication technology and its application to the library and information centers.

These types of practical oriented skill enhancement programmes are very necessary and prolific for the library staff, which can introduce them with advanced technology and refresh and retrained them with various technological skills.

The rapid developments of information and communication system has brought a revolutionary change in the organization and management of information. In present electronic information age, where information is treated as an economic resource, a marketable commodity and as a social wealth, the librarians are to play an active and important role in the process of information communication system. The libraries are facing new challenges, new competitors, new demands, new expectations and a variety of information services from users. They are now to be more acquainted with the skill of handling new technologies related to collections processing and dissemination of information for working in the borderless digital library environment, besides gaining the professional knowledge in library and information science, the library professionals should have the knowledge of information technology and its application in library operations and services, both in theoretical as well as practical level with the changing role and responsibility of the librarians their professional identity is also changed they are now known as information officer, information broker, cybernarian etc. As the technological revolution has given raise to some innovation and innovative ideas including artificial intelligence, bar code technology, speech synthesis, digital signature etc. The library professionals must become a continuous learner. In a digital web environment, the librarians gets the opportunity to search different websites, e-journals etc. as well as to share ideas among the same professionals groups using e-mails, audio/video conferences etc. moreover, in the borderless library the librarian is to offer continuous consultancy services to its users throughout the web.

Conclusion

So in digital environment, we need libraries and librarians more than even before in order to make effective use of the information that is available, we need them to identity quality information, to preserve it, organize it and to make it accessible by cataloguing and data management. The librarians and library and information professionals are required to acquire such knowledge and skills as the library is one of the highly it influenced service profession and the empowerment of library and information professionals with it skills is aimed at providing services that are expected of from the clientele in the new environment.

Development of information technology is playing a crucial role in restructuring of the libraries. Shift from human deponent operations to machine dependency, mechanization to knowledge processing stand alone system to network computing, local LAN to wireless access protocol systems. Document centered information to user centered information, print media to electronic media, data capture methods, human to machine oriented, library automating to web enabled services, WAN access, and online information retrieval to CDROM databases to internet. These prolonged shift in application of innovative IT to library and information profession can be attributed to the changes emanated in the last decades.

However, borderless library is becoming available to the users community at their door step but this beneficial technological application in a library institution will be fully successful only when there is a close co-ordination between the IT and the human resource. Because it is mandatory for any institution that before implementation of any information technology, it should be ensured that the employees possess the required skills and competence. In the digital borderless library environment the LIS professionals have to face different kinds of tuff challenges, so there is the maximum need of skills enhancement programmes for the library staff. Besides academic institution, other organizations may also help in enhancing the skills of the library professionals. Along with the potential working skills, the library professionals should also have the positive attitude to work for the benefit of institutions for achieving its goals.

Practical oriented skill enhancement programmes are very necessary and prolific for the library professionals, which can introduce them with advanced technology and refresh and retrained them with various technological skills.

Digital age has brought a tremendous change in the way information is stored and accessed. This has brought about a change in the concept of librarian, their collection and services.

In this respect library professionals have great responsibility not only to provide information available in all kinds of sources for productive information work but also work with the policy makers to ensure that an information literate society becomes a governmental priority. The library professionals and the web technologists have to play the key role for its proper implementation and developing high bandwidth, high quality communication technology which can reach to every sectors of the society. As the national knowledge commission, under the chairmanship of Sam Pitrodar has stated that for establishing a knowledge dependable society, importance should be given in greater participation and more access to knowledge across all sector of the society. The role of library professionals has been changed in e-age or e-resource library environment it is therefore pertinent on the part of library professionals to acquire new dimentations and skills required for management of the digital libraries and e-age library.

References

1. **Chopra H.S.**, "*Library Information Technology in Modern Era: Libraries and Librarians*" in New Millennium Commonwealth Publishers, New Delhi, 1999.
2. **Creth, Sheiles D.** (1996). "*The Electronic Library : Slouching Toward the Future or Creating a New Information Environment*". Paper Presented at Follett Lecture Series. Available at http://www.ukoin.ac.uk/services/papers/follett/creth/paper.html
3. **DHAWAN, S.M.** "*Role of Libraries in the Emerging Electronic Information Era.*" In : Library and Information Studies in Cyber Age; Essays in Honour of Prof. J.L. Sardana. Edited by S.M. Dhawan, Delhi: Author Press, 2004, pp. 500-510.

4. **Kanjilal, Uma** (2004). "*Education and Training for Digital Libraries : Model for Web Enhance Continuing Education Programme in International Conferences on Digital Libraries*". New Delhi Feb. 2004, p. 620-635.

5. **Lancaster, F.W.** (1997). "*Artificial Intelligence and Export System Technologies*". Prospects : In libraries for the New Millennium. Implication for managers. Library Association Publishing, London.

7. **Nageswara Rao, K. and Babu, K.H.** (2001). "*Role of Librarian in Internet and World Wide Web Environment, Informing Science,* 4(1): 25-34.

8. http://www.growingupdigital.co

9. http://www.voidspace.org.uk

10. www.kar.nic.in/publibsclhtm (website of karnataka state public libraries)

25

Open Access Electronic Journals in the Field of Agricultural Sciences

Pravin Kumar Singh,
Sanjiv Saraf & A.K. Rai

Abstract

Since past few years free online information sources like e-journals, e-books, e-databases have increased considerably. Amidst the furor of serials price increases, there is a steadily growing segment of the serials landscape with prices that can't be beat, that is to say, free. Many scholarly journals are freely available in electronic format (no subscription fee or membership required). This article provides information about a fairly comprehensive list of open access electronic journals along with their links in the field of agriculture science and its impacts on libraries.

Keywords: *Open access, Electronic journals, E-journals, Agricultural sciences, Free electronic journals, E – Agriculture Journals, Internet.*

Introduction

Human being while traveling through Pastoral, Agrarian and Industrial Societies has now reached 'Information Society'. With the advent of Information Communication Technology, there are now drastic changes almost in every sphere of life. Within a decade, it has absolutely changed the working of the traditional libraries and uses traditional print resources to electronic resources.

Agricultural science is a broad multidisciplinary field that encompasses the parts of exact, natural, economic and social sciences that are used in the practice and understanding of agriculture. Agriculture is the set of activities that transform the environment for the production of animals and plants for human use. Agriculture concerns techniques, including the application of agronomic research.

Information resources are unlimited but the budget of the institution is limited, with the increase in the price of journals, no library in the world however rich it may be, is not capable of purchasing all the journals produced in the world. Libraries are important resources for individual as well as for communities and organizations. The electronic revolution, specifically, Internet is narrowing the information gap. The power of web technology is enabling the generator of information to disseminate their creativity at low cost and high speed. Internet is the gateway for libraries and information centers to enter the Electronic Information Era and is providing the information, generated by different organizations, institutions, research centers and individuals all over the world.

Electronic Journals

Different people might have a different impression or understanding of the term "electronic journals". Electronic journals are often referred to interchangeably as "electronic publishing", "electronic serials", "online journals" and "electronic periodicals". There are certain intrinsic factors that make these terms interrelated or equivalent. With new developments in technology, the distinctions are not easily drawn. Nevertheless, in order for this literature review to be meaningful, a clear terminology needs to be established. According to *Webster's Third New International Dictionary of English Language*, "journal" is defined as "a periodical publication, especially dealing with matters of current interest; often used of official or semiofficial publications of special groups." In AACR2R, the definition of a serial is "a publication in any medium issued in successive parts bearing numerical or chronological designation and intended to be continued indefinitely." Obviously, these concepts do not fit perfectly to define electronic journals.

Scholarly journals available in electronic format are fast becoming the rule and not the exception. In their early development, there had been a tradition of creating/offering free access to electronic journals, perhaps taking advantage of the lower production costs of electronic distribution. Many journals are freely available in electronic format.

Open Access Electronic Journals in Agricultural Sciences

Open access journals are scholarly journals that are available online to the reader "without financial, legal, or technical barriers other than those inseparable from gaining access to the internet itself".

Online information sources are referred to be free when access to the sources is not dependent on a subscription or membership in an organization or a publisher. Among the freely available online information sources, which are identified and compiled by the authors along with their URL, following are the important ones.

Apart from the above mentioned list some of the important portals which are available for free access of electronic journals in the field of agricultural sciences are:

1. AGORA : http://www.aginternetwork.org/
2. AgNIC : http://www.agnic.org
3. DOAJ : http://www.doaj.org
4. High Wire Press : http://highwire.stanford.edu/
5. HINARI : http://www.who.int/hinari/
6. INFOMINE : http://infomine.ucr.edu/

List of Some useful Open Access Electronic Journals and Websites in Agricultural Sciences

S.N.	Name of Journals	URL
1.	Acta Agriculturae Slovenica	http://aas.bf.uni-lj.si/index-en.htm
2.	Acta Ichthyologica et Piscatoria	http://www.aiep.pl/index.html
3.	Aestimum	http://epress.unifi.it/riviste/CMpro-v-p-52.html
4.	African Journal of Agricultural Research	http://www.academicjournals.org/ajar/
5.	African Journal of Food, Agriculture, Nutrition and Development	http://www.bioline.org.br/nd
6.	Agriculture 21	http://www.fao.org/ag/magazine/default.htm
7.	Agricultura Técnica	http://www.scielo.cl/scielo.php?script=sci_serial&pid=0365-2807&lng=en&nrm=iso
8.	Agronomy Research	http://www.eau.ee/~agronomy/
9.	Agricultural Research	http://purl.access.gpo.gov/GPO/LPS192
10.	Agricultural Engineering International : the CIGR Ejournal	http://cigr-ejournal.tamu.edu/
11.	Alaska Fishery Research Bulletin	http://www.adfg.state.ak.us/pubs/afrb/afrbhome.php
12.	Animal Research	http://www.edpsciences.org/journal
13.	Annals of Botany	http://aob.oxfordjournals.org/archive

Contd....

S.N.	Name of Journals	URL
14.	Annals of Agricultural and Environmental Medicine	http://www.aaem.pl/index.html
15.	Applied and Environmental Microbiology	http://aem.asm.org/contents-by-date.0.shtml
16.	Applied Ecology and Environmental Research	http://www.ecology.kee.hu/
17.	Applied Entomology and Zoology	http://www.jstage.jst.go.jp/browse/aez
18.	Australian Journal of Basic and Applied Sciences	http://www.insinet.net/ajbas.html
19.	Basic and Applied Mycology	http://www.doaj.org/goto/www.bio.unipd.it/bam/
20.	Bioinformatics	http://bioinformatics.oxfordjournals.org/archive/
21.	Biotechnology and Development Monitor	http://www.biotech-monitor.nl/new/index.php?link=publications
22.	BMC Biotechnology	http://www.biomedcentral.com/bmcbiotechnol/archive/
23.	BMC Plant Biology	http://www.pubmedcentral.nih.gov/tocrender.fcgi?action=archive&journal=59
24.	BMC Veterinary Research	http://www.pubmedcentral.nih.gov/tocrender.fcgi?action=archive&journal=327
25.	Breeding Research	http://www.jstage.jst.go.jp/browse/jsbbr
26.	Briefings in Bioinformatics	http://bib.oxfordjournals.org/archive/
27.	Canadian Journal of Veterinary Research	http://www.pubmedcentral.nih.gov/tocrender.fcgi?action=archive&journal=133

Contd....

S.N.	Name of Journals	URL
28.	Canadian Veterinary Journal	http://www.pubmedcentral.nih.gov/tocrender.fcgi?action=archive&journal=202
29.	California Agriculture	http://calag.ucop.edu/
30.	Electronic Journal of Environmental, Agricultural and Food Chemistry	http://www.ejpau.media.pl/
31.	Electronic Journal of Polish Agricultural Universities	http://www.ejpau.media.pl/
32.	Experimental Animals	http://www.jstage.jst.go.jp/browse/expanim
33.	Fishery Bulletin	http://fishbull.noaa.gov/index.html
34.	Florida Entomologist	http://www.fcla.edu/FlaEnt/)
35.	Garden and forest	http://quod.lib.umich.edu/cgi/t/text/text-idx?page=browse;c=gandf
36.	Genetics	http://www.genetics.org/contents-by-date.0.shtml
37.	Genetics, Selection, Evolution	http://www.gse-journal.org/
38.	Genome Research	http://www.genome.org/contents-by-date.0.shtml
39.	International Journal of Applied Research in Veterinary Medicine	http://www.jarvm.com/
40.	International Journal of Botany	http://ansijournals.com/3/c4p.php?id=1&theme=3&jid=ijb
41.	International Journal of Dairy Science	http://www.academicjournals.net/c4p.php?j_id=ijds

Contd....

S.N.	Name of Journals	URL
42.	International Journal of Poultry Science	http://www.pjbs.org/ijps/ijps.htm
43.	International Journal of Sociology of Agriculture and Food	http://www.csafe.org.nz/ijsaf/
44.	Internet Journal of Veterinary Medicine	http://www.ispub.com/ostia/index.php?xmlFilePath=journals/ijvm/front.xml
45.	Israel Journal of Veterinary Medicine	http://isrvma.org/journal.htm
46.	Japan Agricultural Research Quarterly	http://ss.jircas.affrc.go.jp/english/publication/jarq/index.html
47.	Japanese Journal of Crop Science	http://www.jstage.jst.go.jp/browse/jcs
48.	Journal of Agronomy	http://ansijournals.com/3/c4p.php?id=1&theme=3&jid=ja
49.	Journal of Animal and Veterinary Advances	http://www.medwellonline.net/java/fp.html
50.	Journal of Extension	http://www.joe.org/
51.	Journal of Insect Science	http://www.insectscience.org/
52.	Journal of Biosciences	http://www.ias.ac.in/jbiosci/index.html
53.	Journal of Central European Agriculture	http://www.agr.hr/jcea/
54.	Journal of Cotton Science	http://www.cotton.org/journal/
55.	Journal of Dairy Science	http://jds.fass.org/contents-by-date.0.shtml
56.	Journal of Equine Science	http://www.jstage.jst.go.jp/browse/jes/-char/en

Contd....

S.N.	Name of Journals	URL
57.	Journal of Genetics	http://www.ias.ac.in/jgenet/index.html
58.	Journal of Pesticide Science	http://www.jstage.jst.go.jp/browse/jpestics
59.	Journal of Poultry Science	http://www.jstage.jst.go.jp/browse/jpsa
60.	Journal of the International Society of Sports Nutrition	http://www.sportsnutritionsociety.org/site/journal/journalindex.php
61.	Journal of Veterinary Medical Science	http://www.jstage.jst.go.jp/browse/jvms/-char/en
62.	Journal of Veterinary Science	http://www.vetsci.org
63.	Journal of Heredity	http://jhered.oxfordjournals.org/archive/
64.	Livestock Research for Rural Development	http://www.cipav.org.co/lrrd/
65.	Pakistan Journal of Nutrition	http://www.pjbs.org/pjnonline/index.htm
66.	Plant Biotechnology	http://www.jstage.jst.go.jp/browse/plantbiotechnology
67.	Plant Cell	http://www.pubmedcentral.nih.gov/tocrender.fcgi?action=archive&journal=95
68.	Plant Physiology	http://www.plantphysiol.org/contents-by-date.0.shtml
69.	Plant Methods	http://www.plantmethods.com/articles/browse.asp
70.	Plant Production Science	http://www.jstage.jst.go.jp/browse/pps
71.	Research Journal of Agriculture and Biological Sciences	http://www.insinet.net/rjabs.html

Contd....

S.N.	Name of Journals	URL
72.	Revista de la Facultad de Agronomia	http://www.revfacagronluz.org.ve/
73.	Scientia Agraria	http://calvados.c3sl.ufpr.br/ojs2/index.php/agraria
74.	Scientia Marina	http://www.icm.csic.es/scimar/sci_index.html
75.	Sheep and Goat Research Journal	http://www.sheepusa.org/index.phtml?page=site/researchopener&nav_id=601e0a31bbf6a0ef56f1f0591aa0dc78
76.	South African Journal of Animal Science	http://www.sasas.co.za
77.	Tropicultura	http://www.bib.fsagx.ac.be/tropicultura/
78.	Turkish Journal of Agriculture and Forestry Sciences	http://journals.tubitak.gov.tr/agriculture/index.php
79.	Turkish Journal of Fisheries and Aquatic Sciences	http://www.trjfas.org/
80.	Turkish Journal of Veterinary and Animal Sciences	http://journals.tubitak.gov.tr/veterinary/index.php
81.	World Journal of Agricultural Sciences	http://www.idosi.org/wjas/wjas.htm

Impact on Libraries

Freely available online information sources made a huge impact on the library services and collection of the libraries and information centers.

- Free electronic journals reduces the financial problems of the library & information centers as there is no financial involvement to procure such freely available sources which resides in remote server, except the nominal cost of Internet connectivity and surfing. At the same time it reduces the cost of maintenance such as preservation and binding etc.
- Freely available e-books, e-journals, online database are the additional sources of any library collection development.
- Once the links are created/compiled in the web page of the institution or in any particular site the technical processing of such huge documents like e-books, e-journals etc is not necessary. The users can access their required document and can download the same.

Conclusion

The challenge of the libraries in India are facing with is not a paperless society but to maintain, nurture and optimize the resources of the libraries with the help of new technology. Since this is the age of electronic era, mutual exchange of information and resource sharing will be the objective of our professional ethics and it has to be nourished in future for growth and healthy development of our library and information centers. The escalation in the cost of journals has made the libraries and information centers handicapped in purchasing journals. In this situation of limited factors like money, space, manpower and materials etc. there is a need for Inter library cooperation and resource sharing. It is duty of the library and information professionals to find out, which of the sources are useful to their users and to integrate their resources and provide better service to the users. There is no doubt that this type of compilation of the website of freely available online journals will be more relevant and effective to save the money and manpower of Library and Information Center also helpful to the research scientists who are engaged in the field of agriculture and related discipline.

References

1. **Ghosh, T.B.** (2002). "*Freely Available On–Line Electronic Information Sources on Internet and their Impact on Libraries and Information Centers*". CALIBER-2002, 14-16 February, 2002, Jaipur, INFLIBNET Centre, Ahmedabad .
2. **Ghosh ,T.B.** (2003). "*Free Online Electronic Information Resources on Applied Science and Technology*". Paper presented in proceedings 48th All India Library Conference (ILA) on Electronic Information Environment and Library Services: A Contemporary Paradigm, pp. 36-45, NIMHANS, Bangalore, India.
3. http://en.wikipedia.org/wiki/
4. http://www.library.ucsb.edu/istl/00-summer/refereed.html
5. http://www.library.unr.edu/ejournals/free.aspx
6. http://www.aacr2.org/
7. http://www.websters-online-dictionary.org/
8. http://www.britannica.com/
9. http://www.doaj.org/
10. **Mukherjee, Bhaskar** (2008). "*Open Access Scholarly Publishing in Library and Information Science*", Annals of Library and Information Studies Vol. (55), pp. 212-223.
11. **Shukla, Paraj and Singh, Anand P.** (2009). "*Open Access Initiatives for Agricultural Information Transfer Systems in India*". Paper presented in World Library and Information Congress: 75th IFLA General Conference and Council 23-27 August 2009, Milan, Italy.
12. **Shukla, Ramakant K. and Yadav, Manju** (2007). "*Free Online Journals in the Field of Social Science: An Analytical Study*". Paper presented in proceedings EMPI Digital Library Conference 2007, New Delhi, India.
13. **Taranum, Ayesha** (2001). "*Free Electronic Scholarly Journals : A Case Study*". Paper presented in CALIBER-2001, pp. 271-276, University of Pune, INFLIBNET Centre, Ahmedabad .

26

Issues on Preservation of Digital Information in Libraries

P. Ganesan

Abstract

Preservation of digital information is very much needed for ensuring the availability and long-term access of the same. Lot of digital information is being lost, because of lack of knowledge on preservation policies and techniques to be used. In this paper, an attempt has been made to study about preservation, need for digital preservation and issues in preservation.

Introduction

Preservation is one of the foremost important activities for both traditional and electronic library. A preservation technique for long-term and creating Meta data for organising and retrieving the relevant information has to be considered at the earlier stage i.e. before digitization begins. Deciding the preservation strategies at the earlier will have high impact on digital preservation and will lead to the right direction in the preservation process. Preserving the digital information is a challenging job for libraries, because the growth of information is alarming and much of the information is being already lost, because of lack of knowledge on preservation policies and techniques to be used. Revolutionary change has occurred in the preservation due to the invention of digital storage media on various types and the other information technological advancements like proliferation of high-end personal computers, scanners and high speed networks. It will be the duty of the library professionals to preserve all the library materials by using appropriate technological tools, for ensuring the long-term preservation and access of the digital information. In this paper, an attempt has been made to study about the preservation of digital information, need for the digital preservation, and issues involved.

What is Preservation?

Digital preservation is taking steps to ensure the longevity of digital documents, either 'born digital information' or converting the printed materials into digital materials.

Why Digital Preservation?

In the library, two types of digital information has to be preserved namely 'born digital' and printed materials, which are to be digitized. In the digitization process, apart from preserving the 'born digital information', it is also necessary to preserve the digitised printed documents also. There should be proper digitization polices and strategies for both type of information and special care should be taken for born digital, since most of the born digital information is being lost due to lack of planning for long-term preservation and access techniques. The major reasons for digital preservation are for ensuring the longevity, security, simplifying the information management and easy access. In this digital environment, preservation is becoming increasingly important and urgent for libraries and plays key role in preserving the digital information for future generations.

There are three common approaches to digital preservation:

- ***Migration*** : Migration describes the process of copying content from one format (such as a CD-ROM) onto a newer format (such as a solid state flash drive).
- ***Refreshment*** : A related process is refreshment. Refreshment involves copying data onto a newer. example

of the same format (such as from an old CD-ROM to a new CD-ROM).

- ***Emulation*** : Emulation is a more involved process of accessing data on a system other than the one it was made for. Commonly, this will be because an original system is no longer available.

Issues in Preserving Digital Information

Unlike the non-digital formats like books, journals, theses, dissertations, manuscripts or microfilms, the digital formats requires combination of softwares, hardwares and networks for creating and preserving digital information. There is a thread in softwares and hardwares that these technological gadgets can become obsolete within short period. Ensuring long-term access and preservation requires current technological gadgets. Apart from this, there are some difficulties, which include technical, social and legal issues listed below.

- Technological obsolescence is the major problem in preserving the digital materials. Most of the companies started innovating new computers and softwares with greater storage capacity, processing speed for lower cost. The devices are outdated with short period due to innovation and market competition. Because of these advanced technological changes, the materials created in digital form and materials converted from retrospective to digital form are equally vulnerable to technological obsolescence.
- The preservation of digital information is continuous one and initial cost for the digitization will be very high and additional cost should be ensured for migrating the data from one medium to another medium.
- Lack of awareness on preservation strategies, intellectual property rights, migration form to new technology are some of the problems.
- Absence of established standards, protocols, and standard method for preservation.

- The complexity of integration of texts, images, audio, video and their dependence on software.
- Lack of experts and fear of changing or moving to the new changes.

Move to Digital Preservation Now and Forever

The ultimate goal of the digital information is to ensure the long-term access. Though there are number of problems in preserving the digital information as mentioned above, it is very much needed and to be ensured the issues involved in preservation. Some of the major issues like longevity of hardwares, softwares, migration etc to be kept in mind before planning for preservation. Recent experiment shows that everything preserved today is based on the widely accepted standards and requires minimal expense and manpower that integrate the current and future digital systems and will remain for so many years. Hence the long term availability of digital information is required and action must be taken accordingly to make use of the digital information now and then.

Conclusion

The preservation of digital information will vary depending upon the data. The major issues of digital information are ensuring its availability and question of affordability. Hence to ensure the availability, strong preservation policy should be framed and proper planning is needed as discussed.

References

1. **Margaret Hedstrom**, "*Digital Preservation: a Time Bomb for Digital Libraries*", 2002, University of Michigan (http://www.uky.edu/~kiernan/DL/hedstrom.html).
2. **Alison Bullock**, "*Preservation of Digital Information : Issues and Current Status*", April 22, 1999, National Library of Canada.
3. **Michael W. Gilbert**, "*Digital Media Life Expectancy and Care*", 2003, UMASS.
4. **Ganesan, P and Varatharajan, N.**, "*Application of Scanners and Its Role in Digitization of Printed Materials*", Calicut University, National Seminar on Library Co-operation in networked world, 2001.

27

Paradigm Shits in the Library and Information Services : Past and Present Schema

Muzamil Mushtaq &
Sheikh Mohd Imran

Abstract

As the new screen of information environment have stirred the existing systems of knowledge structure and architecture into a transit mode and the very alignment, bent and slanting towards this fascinating environment is imperative for one and all. Same is crystallised and cleared about the Libraries and Information Centers which are struggling (positively) hard to undermine the influence and implication of the new ways of creation, codification, collation, and distribution of the unoccupied content stretched across the vast reservoirs of the knowledge domains of life. The present paper weigh ups the renovation of a traditional library into a modern library made by comparing the Schemas, pre-requisites and practices of the modes and methods used earlier with the Schemas, pre-requisites and practices being followed today.

Approach/Methodology/Design: *The paper is conceptual visualising the various facets of information handling existed in the yesterday's Libraries and Information Centers and coinciding its services with the today's custom-made new born Information Centers in the light of changing state of affairs.*

Keywords: *Paradigm Shifts; Traditional Libraries; Modern Libraries; ICT Impact; Tools and Techniques of new librarianship.*

Introduction

Libraries as we say are the warehouses and databanks of information which has been indexed and cataloged to make retrieval of information more paced, vibrant, easier and efficient. As various sectors of academic communities are harnessing technology to enhance academic programs, the library, at the center of research and scholarship, is also leveraging the new technology capabilities to provide new and enhanced services. The pace of technology innovation in libraries has steadily accelerated over the past decade. Initially technology tools were being applied to the same fundamental library service paradigms to make the work more efficient, but now library work itself is beginning to change, with technological innovation leading to design of new services for users **(Moyo, 2004a).** Traditionally, libraries and the librarians had defined collection development policy about the type of collection to be maintained, but now it is more strategic and dimensional with greater amount of free handedness. Originated from simple tools and procedures to handle the limited information, but with the advent of internet, World Wide Web, ICT applications/usage, library automation, computer and library networks, sophisticated advances in databases, web and libraries 2.0, blogs, wikis, RSS feeds, videoconferencing, etc. the libraries are in a transitory stage to either align the existing services or shed off them and go for new innovative means and methods of information coordination. The impacts on library services will have implication not only for internet access, training, and services, but for expectations and perceptions of libraries by patrons and communities, the reliance of governments on libraries, and many other central library activities as well. It is imperative to think libraries, both to provide services with approaches and strategies and to understand changes to uses and expectations of libraries in such times. From the plethora of information available on the Internet and the World Wide Web, what to be selected and maintained is a significant subject. The emergence of e-publishing, self-publishing, open access publishing, etc. further complicated the collection building policies **(Varalakskhmi, 2009a).** Today the libraries provide and sustain public access internet services and resources that meet patron, community, and government access needs and expectations, serve as key technology and internet–based resource/service training centers in their communities and install, maintain, and upgrade the technology infrastructure required to provide public access internet services and resources **(Bertot, et.al, 2009).** Libraries have long been committed to the serving the interests-be they educational, recreational, economic, or governmental, of all elements of their communities.

Library Services Shifting : Driving Factors

The seepage of the user's community and their new ways of seeking the information and constant innovation in the ICT environment force the existing preservators of the knowledge to shift/mould their services to the changing module and customers. The speed of present day microprocessors, decreasing size of storage media and moving towards nano storage, global access to Internet, increasing speed of search engines in searching and retrieving information, efficient computing devices have influenced the quality and quantity of data which can be accessed from anywhere at any time. Neil McLean, the librarian of Macquarie University, stated in a keynote address to a UK/Australian seminar on cooperation in London in July 1997 that the new service paradigms will include the following **(as cited by Moyo, 2004b)**:

- The capacity to influence both the form and makeup of information consumed;

- The personalization of information/ communication services;
- Location should not be a distinctive barrier to access;
- Transparent access is expected to a range of information resources;
- The ability to access a number of remote information sources at the (same) time;
- The ability to mix different media in real time;
- The provision of a choice in suppliers; and
- The ability to interact with other colleagues in a collaborative networked environment.

In the new environment, library patrons are increasingly working more with multimedia information, such as images, video, audio, than with purely textual information. Moreover, they are increasingly conducting their library research in the Web environment. Many electronic library services are now geared to meet this new working environment. **Libner (n.d)** in his vision for the 2012 Library has described: "we move from a single library to a network of libraries; from one collection to distributed collections; from the catalog interface to multiple interfaces; from books and journals to information fields and streams encompassing traditional and non-traditional forms of scholarly communication. These include such diverse forms and genres as preprints, traditional publications, informal commentary, data sets, software applications, maps, video clips, listserv archives, and web pages - all accessible, at least in principle, anytime and anywhere." Some of the factors, which contributed to the transition traditional Libraries into next generation Libraries, are:

- The way people access information – from traditional library visit to network access from place of his convenience.
- With the convergence of computer technology and telecommunication, the digital media has overtaken print

media. There is rapid movement of technologies towards nano storage and real-time access.

- The transition of libraries from information storage and retrieval to information facilitators and aggregators.
- Instead of ownership of information the present generation gives importance to access to information.
- From information precious library environment to vast, readily and to some extent freely available information society.
- Libraries are going for consortia rather than standalone which made possible to have access to more number of journals in science and arts.
- Availability of full-text databases through digitalisation helped to have open access to information rather than closed access.
- Collaboration made it possible for global access to information. Now information can be accessed 24x7 instead of fixed hours.
- The changing methods of information search, today it is text or touch, but next generation search technology is ought to include reflectivity, density, tone, speed, and volume **(Moorthy, 2006).**

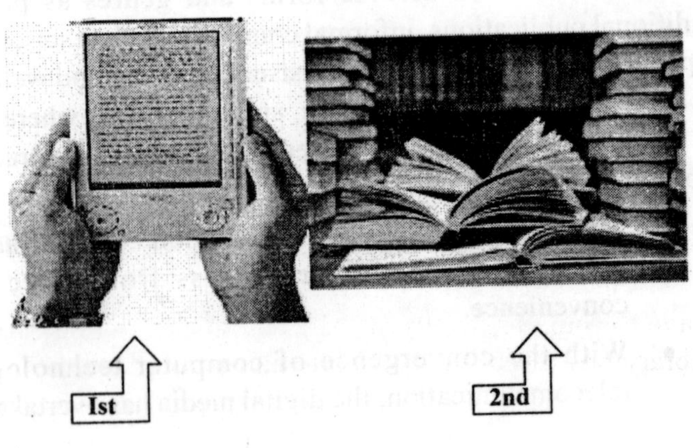

- In this new information environment the user is considered as consumer of knowledge.

The revolution of digital content and networked environment made the libraries boundary less whereby data and information can now be transmitted to all corners of the world, and also can be accessed without geographical restrictions. Now information is just like any other commodity, which can be bought and sold in the market place **(Varalakskhmi, 2009b).**

Preferences

Libraries and Information Centers have to adopt the possibilities of technology application and prepare for new tasks in the process of scholarly communication. The library is fast becoming an information centre providing access to information that is available not only locally but also nationally and worldwide. Major shifts identified in Libraries are:

- from Ownership Model to Access Model
- from Print to Electronic Media
- from Libraries as Archives to Libraries as Access Points
- from Information Collection to Information Analysis and Repackaging.

Libraries were once thought of as buildings with books, periodicals etc in physical form. Now they have evolved into a decentralised network providing services and resources not limited to what they have. Libraries are now in an environment of change characterized by:

(*i*) Greater access to a range of learning resources

(*ii*) Increased speed in acquiring information

(*iii*) Constantly changing technology, and

(*iv*) Lack of standardisation of both hardware and software etc.

This implies changing role and responsibilities and libraries have to find their ways of justifying their future existence. The library has to be highly visible and it should be considered as essential to the teaching, learning and research activities of the university **(Goswami, 2009).**

Service Shifting vis-à-vis Alternate Schema

The start point in libraries is the document procured and the vast array of handling processes to made it readily and easily available to the patrons. The Print monopolism of documents existed in earlier schema of Libraries but now an agglomeration of both Print and Electronic resources which include the following are available.

- Books
- Technical Reports
- Conference Papers
- Patents
- Standards
- Journals
- Policy & Plan Documents
- Staff Profiles
- Video
- E-resources
- Course materials
- Multimedia
- Graphics
- Manuals
- News Clips
- Lectures
- E-mail Archive
- Photographs
- Numerical data

In age olden libraries, the restricted circulation, low level personal reference and information services were the major services being rendered by the custodians of the information and what we are witnessing now is the immense and vast multi-dimensional services of the libraries as can be mentioned below:

- Information products
  - — Formats,
  - — Contents,
  - — Methods of production & delivery of information products.
  - — New business model for use of information products.

— Requires procedural and infrastructural changes and cost implications in Libraries.

- Emergence of Internet as the largest repository of information and knowledge.
- Extinction or significant transformation of some of the conventional information services such as press clippings, contents pages, company information etc.
- Use of new tools and technologies for storage, retrieval and dissemination of information.
- Transformation of role of LIS professional as the subject specialist and end-user gets directly involved in the information work and consequent need for new skills.
- Shift from physical to virtual services that offer convenience of time and location for access to services.
- ICT enabled conventional LIS services

  — Library Catalogue – WebOPAC, Web Browser

  — Information Services- Reference Service, Bibliography, Current Awareness/SDI

  — Document Delivery – Automated

  — Interlibrary loan and union catalogue – Centralized, Distributed, Automated

  — Audio Visual Services – Format, Storage, Access, Transmission, Multimedia

  — Customer Relations and User Education - Communication medium, LIS-serves, Video Conferencing.
- ICT based new LIS services

  — Internet Access

  — Access to web-based resources:

  E-journals, E-books, Course ware, Patents, Standards

  — Subject Gateways – Portals (Bubl, Auslit, AeroInfo),

— Links

— Developing and delivering local resources - Digital Libraries, Repositories, Archives, etc **(Narayana, 2009).**

Comparative Schema of Past and Present

There can be seen two extreme terminals of libraries – one with the traditional time consuming ways of handling the information and other the complete electronic library with virtual interface and in between we have the transitory form of libraries composed of matrix of traditional and modern forms of handling the knowledge reservoirs.

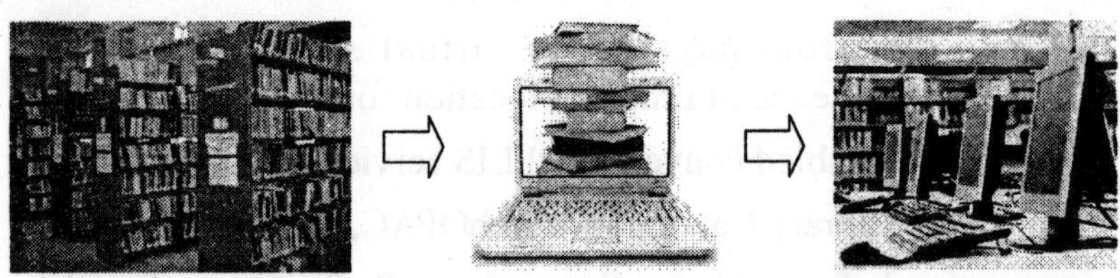

Traditional Screen ***Transitory Screen*** ***Electronic Screen***

But here we have depicted the schema of only the two extremes of past and present Libraries.

Past Schema

- ***Manual Operations*:** The large variety of library housekeeping operations in the traditional libraries were very slow as every operation was done manually.
- ***Time Limits*:** The libraries used to fix for the specified time periods, confined to working hours of the institution and thus halts the very aim of the librarianship.
- ***Hard Access*:** Users were not giving the free hand to search their required documents.
- ***Non Virtual Reference Services*:** There were the personal referencing systems which gives the limited help to trace out the required documents for users.

- ***Unidirectional Mode of Print Resources*:** Non interactive nature of books
- ***Non Electronic Format*:** Documents produced by simple printing press papers
- ***Limited Options to Choose Alternatives*:** The options were less for the users to select the best suits for their demands.
- Websites not available
- All transactions by manual means-there were no automation of library operations
- No search engines

Present Schema

- ***Electronic Operations*:** The mechanisation boosted the services used to be provided by the traditional libraries and thus there is now faster services.
- **24/7 *Service Availability*:** There is now not the restriction of time constraints and users can have the availability of services round the clock.
- ***Easy Access*:** There is greater flexibility now on part of accessing the resources of the library as the open access mode is prominent in present libraries.
- ***Virtual Reference Service Librarian Available Online* 24/7:** The personal referencing systems have now migrated to the online virtual screens.
- ***Easy-to-use Web Resources Permitting Self-service*:** There is a greater easiness in not only accessing the web resources but they provide with the multimedia features which permits a user the perception of self service.
- ***Majority in Electronic Format*:** As the modern library largely depends on electronic equipments, so the maximum of the documents are now in electronic format.
- ***Several Options/alternatives to Choose*:** There are lot of alternative available to the patrons in terms of documents accessing or any other retrieval of information.

- ***A Web Site That Works*:** Modern libraries have a well developed website to cater the needs of the users and other related activities.
- ***Online Library Transactions*:** All the library transactions like library registration, library circulation, request document delivery, interlibrary loan, renew library items, etc. are now done in the electronic pathways.
- A Web site search engine that can find what a user wants.

Conclusion

New roles and new service paradigms and new roles for librarians are emerging as a result of new technologies available for information access, processing and transmission. The modern library users are technology savvy and are constantly moulding towards the digital environment and able to search the Internet. This drifts the essence of libraries to make a progressive shift and today we witness that the libraries can be perceived as facilitators to the Internet, helping surfers access the appropriate information, and develop the right skills for its evaluation and use. The constant shifts in the ways of production, processing, distribution of content and largely by the way of user perception, dynamically and drastically change the screen savers of the libraries and information centers. In future, libraries will reinvent themselves by perfecting the path that they have now begun. They will become more of a fusion of physical and virtual realms. The precise role of the LICs will depend on the organisation structure and knowledge needs. The emphasis in roles of knowledge professional will likely to change in accordance with needs of the user community and the level of technological sophistication. As the technology changes the role of LICs/knowledge centres and knowledge professionals will continue to evolve. At the concluding remarks, I may envelope the paper by the sayings of **Bakker (1999)** "Whatever may be the library environment the primary task has always been - and will remain regardless of changes of technology - to select, stabilise, protect, and provide access to relevant and representative information resources. The collection function, however, is expanding to include a connection function. Selection is moving to an environment in which a multiplicity of media is available. A new facet of resource sharing is the development of joint licensing agreements that permit consortia of libraries to share responsibilities and costs of providing access to electronic resources."

References

1. **Bakker, T.** (1999). "*Resource Sharing in a Virtual Library Environment : User Oriented Collection Management*". Retrieved March 10, 2011, from http://www.sociosite.net/bakker/resovirt.html.
2. **Bertot, J.C., Jaeger, P.T., McClure, C.R., Wright, C.B., & Jensen, E.** (2009). "*Public Libraries and the Internet 2008-2009: Issues, Implications, and Challenges*". *First Monday*, 14, (11). Retrieved March 8, 2011, from http://firstmonday.org/htbin/cgiwrap/bin/ojs/index.php/fm/article/view/2700/2351.
3. **Goswami, P.R.** (2009). "*Academic Librarianship in India: Towards Exploring Strategic Intent and Core Competencies in the Present Era*". ICAL 2009 – Change Management. pp.342-345
4. **Libner, Kelsey. (n.d).** "*Working the Network: A Future for the Academic Library*". Submitted in Response to the call for papers: Visions: The Academic 'Library' in 2012. Retrieved March 10, 2011, from http://alpha.fdu.edu/~marcum/libner.doc
5. **Moorthy, A.L.** (2006). "*Re-orientation of Library Services*". Keynote address delivered at *National Conference on re-orientation of Library Services in India* organised by APLA, Vijayawada, 18-20 August, 2006.
6. **Moyo, L.M.** (2004a, b). "*Electronic Libraries and the Emergence of New Service Paradigms*". *The Electronic Library*, 22 (3), 220-230. DOI 10.1108/02640470410541615
7. **Narayana, P.** (2009). Paradigm Shift in Tools & Technologies in Providing Aerospace Information Services at National Aerospace Laboratories. In Workshop on "*ICT AND ITS USAGE IN Aerospace*

Libraries for Information Services" AR&DB/ARP, 9TH October 2009 CEPTAM, Bangalore, India.

8. **Varalakskhmi, R.S.R.** (2009a, b). Future of Library and Information Centres in Knowledge Society of India: The Expected Role of Knowledge Professionals. "*DESIDOC Journal of Library & Information Technology*", 29, (2), 75-81. Retrieved March 8, 2011, from http://publications.drdo.gov.in/ojs/index.php/djlit/article/view/244/154.

28

Opportunities and Challenges of E-resources

Yashaswi Gyanpuri &
Umesh Kherajani

The concept of E-resources is applicable to every entity without any exception. In the age of information technology the traditional concept of acquiring information is gradually replaced by accessing information on-line. Over the last decade, electronic resources have become increasingly substantial components of any academic library collection. Now it becomes the life line of any library. In this computer age, e-version of book, journals, etc or e-resource in general have become inevitable and hence it is very much needed to convert the printed version in to e-version for future needs. Therefore, knowledge of the different e-resources, developing e-resources and preservation of them has become the need of the hour.

With advances in information technology the size of physical libraries has reduced. It has been possible due to the digitization of information. The digital and electronic information is based on digitized data/information, which has gradually replaced paper-based records. The traditional libraries are becoming hybrid libraries as they are in the process of doing digitization of their documents and moving towards to become digital libraries.

E–resources are available in several forms and types, some of the popular ones are electronic journals, technical specifications, reports, E-books, CD-ROM/DVD resources, Databases etc. E-resources have the characteristic features that they are potentially huge, comprehensive of everything, do not require physical space.

Other important characteristics of E – resources are

1. Feasibility in full content search ability
2. Time, space and cost constraints are not a problem in search strategies.
3. Hyperlinks lead the users quickly to the required information sources.
4. Easy in archiving the content.

Recent advances in the field of information technology contribute significantly to improve the services of libraries. Now-a-days libraries are not only seen with printed document and non-print document but also with computers. The impact of technologies such as CD-ROMs, multimedia, computer networks, Internet, etc. have lead to a paperless society. With the availability of computers, capable of computing at very high speed and having large disc storage space, it is possible to digitize and store information in the form of high quality graphics, color images, voice signal and video clips at a relatively affordable cost.

Growth of E journals in United State

Year	*Electronic Journal*
1998	200
1999	4,400
2000	5,000
2001	7,600
2002	8,600
2010	> 50,000

The advantage of the electronic resources is ubiquity – many users can simultaneous access a single electronic copy from many locations. Copies can be delivered with electronic speed, and it would be possible to reformat the material as per the reader's preference (e.g. character size). Since readers get a screen display of the object, rather than a physical object, loss rates by theft are eliminated. Digital storage also permits libraries to expand the range of material, they can provide to their Users since audio cassette tapes and records cannot stand a large number of playing without deterioration, their digital representation (digital audio) can produce a format that is much safer and of better quality. Digital material can also permit access to video tapes and new kinds of multimedia materials that are created only on computers and have no equivalent in any traditional format. The digital information can be copied without error. As a result preservation in a digital world does not depend on having a permanent object and keeping it under guard, but on the ability it makes multiple copies assuming that at least one will survive.

The management of digital resources has never been an easy process. The rapid expansion of digital resources compounded with changing formats and sales models in the short life of Web-based delivery systems has particularly made the management process more complex. Further, from the very beginning, details relating to purchasing, licenses, access, and usage have been kept in adhoc systems built by in-house teams or by the individual librarian needing to organize her workflow. It is hard to recall the days when digital resources played only a minor role in library management discussions. Other challenges that the different forms of E resources have to face are expensive technology, inadequate screen resolutions and non availability of the titles in the right format.

References

1. Madhusudan M, "Use of UGC e-journals by research scholars and students of University of Delhi, Delhi", Library Hi Tech, Vol. 26 No. 3. pp. 369-386.
2. Asefeh,Asemi and Nosrat Riyahiniya. "Awareness and Use of Dgital Resources in the Libraries of Isfahan University of Medical Sciences, Iran." *The Electronic Library*, 25(3), pp316-327.
3. E-resource Management. Available on www.wikepedia.com Visited on 12.04.2011

29

The Five Laws of Library Science and E-resources

Vnod Kumar Singh &
Ashok Kumar Shukla

Abstract

In today's electronic environment, Information and Communication Technology (ICT) is influencing the work of all the organisations and libraries are not an exception to it. Due to ICT, the work of LIS professionals has been changed and literature is available not only in print but also in electronic form. Now-a-days libraries and users are more focus on e-resources. This paper is a sincere attempt to present the overview of e-resources and also applied five laws of library science to e-resources.

Keywords: *Five Laws of Library Science; E-resources*

Introduction

The Information and Communication Technology (ICT) revolution and the advent of the Internet has had drastic and far-reaching impacts on the knowledge and information sector and added a new dimension to information retrieval platforms. It has created an environment where rapid continuous changes have become the norms. Developments in information and communication technologies have a profound impact on every sphere and academic activities. Academic libraries are not an exception for this. It has reduced the library stature from custodian of our literature heritage to being a competitor among many others in the information society changes have been noticed in the academic libraries in professionals, collection and policies. Changes have seen in information seeking behaviour of users. Their preferences have been changed. User satisfaction level has been increasing. Now libraries have been able to provide fast and seamless access of information to its users (National Conference, 2010). In the 21st century most of the library resources are being made available in electronic formats such as e-journals, e-books, e-databases, etc. Libraries are moving from print to e-resources either subscribing individually or through consortia.

E-resources

An e-resource is defined as a resource which requires computer access or any electronic product that delivers a collection of data, be it text referring to full text bases, electronic journals, image collections, other multimedia products and numerical, graphical or time based, as a commercially available title that has been published with an aim to being marketed. These may be delivered on CD ROM, on tape, via Internet and so on (Bajpai, Mal & Bajpai, 2010).

Types of E-resources

E-journals, e-books, e-databases, CDs & DVDs, e-theses and dissertations, e-reference sources, e-zines, e-newspapers, etc. are come under e-resources (Singh, 2010).

- ***E-books***: It is an electronic version of a printed book which can be read on a personal computer or hand-held device designed specifically for this purpose.
- ***E-journals***: It is an electronic version of journals, publications used extensively by the scientific and academic community to disseminate research findings.

- ***E-newspapers***: It is the online accompaniments of established newspapers where news articles and the latest updates are published on the web.
- ***E-zines***: E-zines are equivalent to e-newspapers but published by established print magazine publishers. Many magazine titles such as The Economist, Times, National Geographic and so on, have also established online websites.
- ***E-databases***: E-databases may consist of books, periodicals, reports & theses, etc can be converted to electronic form that allows access through network.
- ***Electronic Thesis and Dissertation***: It is an electronic document that explains the intellectual works or research of a researcher.
- ***CD-ROM***: It has provided new dimension for information for information storage and retrieval. Now-a-days most of the work on e-journals has concentrated on distribution via the Internet and CD-ROM as well.
- ***E-reference Sources***: It is an electronic version of references sources, such as Encyclopaedia, Dictionaries, Biographies, etc.

Advantages of E-resources

The advantages of e-resources are (Singh, 2010): (*i*) Allow remote access (*ii*) Available before print version (*iii*) More than one user can use it simultaneously (*iv*) Provide timely access to documents (*v*) Support different searching capabilities (*vi*) Accommodate unique features such as links to related items (*vii*) Save physical storage space (*viii*) It can support multimedia information (*ix*) It can be environmentally valuable (*x*) It can be saved digitally (*xi*) It eliminates printing and postage cost (*xii*) It can easily merge with alerting service (*xiii*) It provides improved access through full text searching (*xiv*) It can solve the problems of missing issue of any journals (*xv*) Any change in e-resources can be made available quickly (*xvi*) In case of any eventuality, calamity such as fire, the resources can still be in tact (xvii) Easy to convert in different languages and (xviii) Do not require physical processing.

Disadvantages of E-resources

The disadvantages of e-resources are (Singh, 2010): (*i*) Initial high infrastructure and installation cost (*ii*) Need special equipment to access (*iii*) Lack of compatibility among different publishers (*iv*) Hardware and software compatibility issues between publishers and users (*v*) Difficulty inherent in relating to a large amount of data on a screen (*vi*) Causes more concern about copyright (*vii*) Uncertainty of permanent access (*viii*) Issues related to archiving (*ix*) Training (*x*) Plagiarism (*xi*) Excessive printing of documents (*xii*) Bibliographical control (*xiii*) Classification and catalogue (*xiv*) Pricing (*xv*) Formats and (*xvi*) Security.

Five Laws of Library Science Applied to E-resources

The five laws of library science were enunciated in 1928 by the late Dr. S. R. Ranganathan. Five laws of library science are (Ranganathan, 1988):

- Books are for use
- Every Reader his/her Book
- Every Book its Reader
- Save the Time of the Reader
- The Library is a Growing Organism

When we applied five laws of library science to e-resources we get the following laws:

- E-resources are for use
- Every User his/her E-resources
- Every E-resource its User
- Save the Time of the User
- The E-resource is a Growing Organism

Implications of the First Law

- Libraries must provide the list of subscribed and public domain e-resources.

- It must be available for 24×7 in campus.
- Provide infrastructure to better utilisation of e-resources.
- Library staff must be helpful.

Implications of the Second Law

- Subscribed the most relevant and wanted e-resources.
- An adequate and competent team of staff is essential.
- Promoting resource sharing/ library cooperation/ library consortium among libraries.
- Selection of e-resources also based on the preference of users (Research Scholar, Faculty Members, etc.).

Implications of the Third Law

- First provide free access to all users.
- If not, then provide user ID and password to restricted users.
- Provide list of new additions.

Implications of the Fourth Law

- Put all e-resources on a single window of the website or portal.
- All e-resources are properly classified on the basis of subject.
- Bandwidth should be fast.
- Provide manuals or training programs so that they search effectively.
- Provide more effective, timely and cross search functionality.
- Put the information in a right format.
- Interface should be user friendly

3.5. Implications of the Fifth Law

- To add new computers in the library to access the e-resources.

- To classify all the new added e-resources according to subject or interest.
- Not purchase those e-resources which are not used by the users.
- Some space should be available for the newly subscribes e-resources.
- Time to time check the usage of the subscribed e-resources.

4. Conclusion

The five laws of library science constitute the basic philosophy of library science and of librarianship. They help us to find a rationale for everything that we do or should do in a library. They keep us constantly alert to the new methods and practices that we should introduce in order that the library may serve its community better. Due to the importance of e-resources, five laws have been reinterpreted in the changing context of e-resources. The implications of each law clearly indicate their validity and usefulness in the expanding role of e-resources in research and development.

References

1. **Bajpai, R. P., Mal, Bidyut K. & Bajpai, Geetanjali** (2010). "*Use of E-resources Through Consortia: A Boon to Users of Indian University Libraries*". 500-503. <http://crl.du. ac.in/ical09/papers/index_files/ical-85_83_195_2_RV.pdf>. Accessed on March 13, 2011.
2. National Conference on ICT Impact on the Knowledge and Information Management (2010). Brochure. <http://accman.co.in/lc14.pdf>. Accessed on March 13, 2011.
3. **Ranganathan, S.R.** (1988). "*The Five Laws of Library Science*". Bangalore: Sarada Ranganathan Endowment for Library Science.
4. **Singh, Vinod Kumar** (2010). "*E-resources Management through Portal: Problems and Prospects*". In Proceedings of the National Seminar on Knowledge Resource Management 2010 : Information Resources, Services and Practices (pp. 61-71). Mumbai: Tata Memorial Hospital.

30

Print to E-Resource : Challenges and Opportunities

Surjeet Kumar &
Rakesh Pant

Abstract

This paper clearly reflects paradigm shift emerging from print to e-resources and growing at a tremendous speed, the knowledge dependent on technological changes and information explosion. The need for e-resource to users are growing and becoming very essential. The impact of web based e-learning and teaching environment has influenced every facet of library and providing new opportunities and challenges to the library and library professional for involvement in the knowledge based society including electronic and multimedia publishing, internet based services, global networking, web based digital resources etc.

Keywords: *E-resource, Technology Challenges, Future vision of library, e-learning environment.*

Introduction

The word Library, in the present days brings in our mind collection of printed books but in the earlier times there was practically no distinction between a "record room" and a "library".

The libraries have changed with time from mere static storehouse of interested collection of documents to dynamic service centers, serving all professional and non-professional utilizing useful and need-based collection of documents. The medium of information storage has changed from clay tablets to papers and now to electronic and optical media of CD-ROMs.

Traditionally libraries have acquired print-based materials including newspapers, maps, pamphlets, illustrations and many other items as well as books. One of the major developments in library and information systems in the past two decades is the advent and spread of Electronic Information Sources (EIS), services and networks mainly as a result of developments in information and communication technologies. The commonly available electronic information sources mainly, CD-ROMs, online databases, online public access catalogue (OPAC) and the Internet and other networked information sources are competing and in some instances replacing the print based information sources. The EIS provides access to information that might be restricted to the user because of geographical location or finances. They also provide access to current information as these are often updated frequently. Access to information is important to individual scientists, groups of scientists or the academic community and research institution for accomplishment of their programs and research projects.

Thus information resources are changing from printed to electronic format. As a result many libraries are currently experiencing a fast transition from print to electronic format while developing collection of an academic, technical/Special Libraries. The collection development method and practices of acquiring library resources have changed in order to deal with the electronic version publications. The role of library and information professionals also needs to be changed to provide best services to the end users.

Increasingly, information resources are available electronically. The electronic resources acquired by libraries and information centers include:

- Computer software
- Databases
- News feeds
- Daily financial information sources
- Stock market reports
- Multi media works
- Image collections
- Information collections etc.

New developments in information technology infrastructure have significant role in transformation of the knowledge management processes. Technology is making the difference, transforming the process of knowledge creation, acquisition and distribution.

Besides this the traditional scientific information has to be followed which has been classified as:

- Primary literature
- Secondary literature
- Tertiary literature

The 21st century is an era of information explosion. There are large numbers of publications take place and it has become impossible for the human being to scan all of them manually. With the advent of computerized databases, searching may be done in the following modes.

- In housing searching
- Searching by databases producers
- Searching by service providers
- On-line searching
- On-line searching by service providers

Industrial, revolution resulted in the explosion of information in the form of books, journals, technical reports, conference proceedings, and patents etc. with the emergence of the internet we are witnessing explosion of electronic information (e-information). In the Internet age the print form of information is being transformed into e-information and is being made available on World Wide Web (WWW).

Objectives

The objectives of this study are given below:

1. The main objective of this study is to analyze and explore the changing vision and the role of future library according to meet the changes and challenges in the e-resources environment.
2. To define and explain the concepts of e-resource and digital learning environment in libraries.
3. To define the various changes and challenges to converting to print resource to e-resource.
4. To define the benefit and opportunities using the new technology.

Trends and Challenges before the Future Library in the E-Learning Environment

The first and foremost challenge before the library professionals to face the future academic needs of the user in the e-learning environment is to provide electronic access to all relevant information and integrate it on networks across the world. The second challenge is to create a new physical library premises with computer network facilities, abandoning the old concept of library as a storehouse, and, the third challenge to future library professionals is to develop new standards and skills for the library profession to meet the user needs in a proactive way. In this e-learning and e-publishing environment, electronic reference services and other support services with various expertise and digital repositories are becoming a must.

The most pressing and pervasive issues and challenges that the library and information science professionals face in the present digital era for providing digital information service to the knowledge society are:

(*i*) New generation of learners

(*ii*) Copyright

(*iii*) Privacy/Confidentiality

(*iv*) Online/Virtual crimes and Security

(*v*) Technology challenges

(*vi*) Manpower

(*vii*) Collection of digital e-resources

(*viii*) Organizational Structure

(*ix*) Preservation / archiving of digital e-resources

(*x*) Lack of clarity in vision

(*xi*) Impact of We-based e-Learning Systems

The New Generation of Learners

Today's students are grown up with latest information and communication technologies. They are coming to higher education with aptitude, knowledge and expectations that have been shaped by the use of the Internet, digital media, and portable communication technologies. Students often begin their search for information with Google or similar commercial or social search engines. The academic library professional must develop a virtual electronic learning system to enhance the student's knowledge and to accommodate an increasingly diverse group of users.

Copyright

An important issue that the present day library professionals are facing in providing electronic/digital information service is the large scale of piracy of software and plagiarism. The cost and timeliness in retrieving the information are also considered. When negotiating access with a publisher, the librarian must agree to certain restrictions on photocopying or distribution of electronic materials. Despite copyright notices and efforts to educate employees and users about intellectual property rights, electronic publications can be easily forwarded to people outside the licensed user group. The library is responsible for maintaining the awareness of all users about copyright issues.

Privacy/confidentiality

Maintaining privacy and confidentiality is another problem in accessing online information. To control pirating of software, copying or downloading all the contents of any e-resource at a time, right to obtain information and right to withhold or ban the access is essential and so there is a delicate challenge between privacy and rights to information. Now a day almost all the users are having their own e-mail accounts and they are often sending and receiving important information and even secret programmes and databases through e-mail itself and storing them for future usage. So maintaining privacy from e-mails is a great issue. Protecting one network from another to maintain confidentiality of information is another problem in securing databases on Internet and Intranet.

Online/Virtual Crimes and Security

"Privacy and security are two sides of the same coin," said Kurtz. "If we can improve Web security, we will be able to have a positive impact on privacy as well." Presently, Web/cyber crimes have become a common threat on internet. To overcome this issue, compulsory Virus Proof procedures should be adopted while downloading e-information from any other system. To secure the system from viruses, the databases can be modified by hacker proof procedures. Separate login and password systems are to be compulsorily adapted to the Network systems. In the LAN environment, the real danger is the gradual erosion of individual liberties through the automation, integration, and interconnection of many small, separate recordkeeping systems, each of which alone may seem innocuous, and wholly justifiable. To overcome the above database security problems and issues, it is essential to install a database security software or firewall technology like

Norton Anti-virus software and IBM e-network Firewall technology to protect the databases.

Technology Challenges

Technology provides challenges to access information. The ALA's 1995 Code of Ethics clearly states that everyone should have access to information. The recent explosion of information available on the Internet presents challenges to the traditional American Library Association (ALA) code of ethics that is taught in library school. Librarians make ethical decisions every day on the basis of the culture of their organizations. Some organizations limit access to particular levels of employees by requiring a username and password; others may institute behind-the-scenes filtering software or restrictive policies for providing access to the entire Internet. Because these steps challenge the very essence of librarianship, the librarian must step in and voice concern for the patron's rights. Establishing well defined access policies will help to clarify who has access to the Internet, under what conditions, for what purposes, and with what restrictions. Policies should consider how to integrate the new technology and how its use reflects the objectives and values of the library.

Manpower Issues

Lack of skilled manpower to maintain the e-resources and to provide proper e-information service to the knowledge society is another main problem. Core competencies of library staff are expanding to include technology skills, personal skills, learning and teaching capacity, team skills, commitment to ethics, leadership skills, communication skills, creativity skills, designing and implementing skills etc. Hence library education must be redesigned to meet the new challenges and issues evolving in the knowledge society. Adequately skilled staff should be recruited to meet the increased demands of the knowledge society. With a rapidly changing environment both within and outside the library, staff development programs are crucial to the continued success of the organization.

Organizational Structure

Technology has broken down the rigid hierarchical structure of the organizations which is another important issue in changing the roles of the librarian in the knowledge society. Far from emulating the organization of conventional libraries, the organization and structure of digital libraries, and the division of labour within them, are open to considerable experimentation. For example, as publishers and professional societies disseminate works electronically, they are testing how far their investments should incorporate the full range of library functions, and the digital libraries license content from publishers and professional societies that manage their own repositories.

Collection of e-resources

Collecting the materials and making it available to all current and future users is another core value of librarianship. The challenge is for the librarian to contribute to establish realistic collection-development policies covering acquisition of and provision of access to electronic resources for users now and in the future. With the increase in electronic resources, librarians and libraries are no longer just collecting and caring for print materials. Unlike a print book or a journal, electronic resources cannot be considered a permanent addition to a collection. Payment for a product covered by a license is a payment to use the information product for a period of time that is usually specified in a contract. This payment is not for the outright purchase of the product or for ownership of all the rights to that product. A digitized collection means that libraries share the use of the collections with other institutions, not only locally, but also globally. It is the publisher who dictates how much access will be provided, which issues will be available, and how much that access will cost.

Preservation/Archiving of e-resources

To preserve the e-resources for access would be a contradiction in an electronic environment for librarians, where there is unlimited and continuous access, but performance is not there in such an environment. This leads to the conflict on what is to be preserved and what is to be accessed. If we need to preserve electronic resources/documents, we need to preserve all the software and hardware also to read the documents that we create. Currently, there are two radically different solutions for preserving digital information: migration and emulation. Neither solution is without some risk. Migration may not work for specialized, proprietary formats. It may save the content of a file but lose or diminish the internal relationships or contexts of the information. The second strategy, emulation, assumes future access to multiple data objects. If one or more of the components were missing, this complex environment would most likely fail.

Distribution and archiving through digital repositories will insure that the library has a viable system for sustaining digital content. Digital repositories also will facilitate the long term conversion and preservation of print materials, and create new opportunities to structure learning activities around the content.

Lack of Clarity in Vision

The biggest challenge that the librarians are facing in the knowledge society, seems to be lack of clarity in vision and a general lack of direction. A general vision is needed and the general integrated plan should be shared among the library professional, which should bring unity of purpose. The Library professional should become capacity builders and facilitators to the knowledge society. The vision of the library professionals should emphasize on the quality of services provided to support teaching, research and public service activities, to enable the users to become self sufficient and to make the library both a place and gateway for accessing information within and beyond the walls of the library.

Impact of We-based e-Learning Systems

The emergence of web-based e-learning systems through Internet facility has great impact on every facet of library activities and information services. Library and information professional of the future academic libraries face the following paradigm shifts due to the rapid developments in the ICT and WWW technologies:

1. Transition from procuring and managing print media to electronic media
2. Changes from passive user to active user in the e-literacy environment
3. Concept of web-based networked environment
4. Disseminating information on demand to proactive digital information services
5. Providing information service to facilitating access to e-information service
6. Transition of developing the normal collection to e-resources (e-books and e-journals)
7. Individual works to team works.

Benefits and Opportunities of E-resources

The abode of knowledge is in transition mode from repositories to open access, dramatic and drastic changes in acquisition, process, storage and dissemination of information, harnessing and apt application of versatile technologies, gap between user needs and services rendered and from phobia of ICT developments to justify the inevitable changes required in e-environment for sustainability and future life. Libraries have been significantly transformed with the advent of internet and the ability to provide resources to people who may never visit a physical building, but use resources intensively in their won homes or work places. The unimaginable developments in the information environment such as improved accessibility, interoperability and open access to educational materials has one side facilitated the nature, role and services but on the other side pose a serious challenge to harness the technology and provide state of the art services, otherwise we will be left behind in the transformational phase.

Libraries are changing dramatically by adopting new mans of technology in all activities of print resources to e-resources like printed library card catalogues have been replaced by computerized OPAC system with a variety of web-based graphical user interface (GUI) functions, online accessibility for 24/7, availability of numerous e-databases, e-journals, information resources, services for users. To face the new explosion, libraries will have to meet even more challenges and opportunities to serve students, faculty, staff, scholars and other users, all with much expectation and many more demands triggered by the growth of emerging and cutting edge technologies in academic learning environments. The new roles challenge and opportunities of library in present era can be seen as:

Gateway of Information

A library has to function as a central gateway for library users to access, locate, transform, and utilize information resources in a variety of printed and electronic formats via applications, databases, networks, platforms and systems.

Learning Centre

A library has to provide library users with dynamic equipment, facilities, resources, and services to support their learning activities, which cover assignments, presentations, projects, research papers, reports, etc.

Training Centre

A library shall provided best supporting and training facilities to faculty and instructors for designing, developing, integrating and implementation of various teaching courses, programmes, workshops including support for distance learning programmes.

Publication Centre

A library shall provide library users with computer hardware and software, audio-video equipment and other supporting facilities and peripheral devices to create, design, develop, integrate, publish, and upgrade their various multimedia presentations, projects, web sites blogs, and so on.

To harness the technological paradigms some new services should be focused upon such as :

Library 2.0

The term Library 2.0, first coined by Michael Casey in 2006 on his blog Library Crunch, refers to a number of social and technological changes that are having impact upon libraries, its staff and their clientele, and how they could interact. It is a model for modernized form of library service that reflects a transition within the library world in the way services are delivered to users. The focus is on user-centered change and participation in the creation of content and community. The application of concepts and technologies of Web 2.0 applied to the library services and collections is named as "Library 2.0". It is a concept that personified new generation of library services to meet the present day users' needs and expectations.

Library Digitization

Library digitization is the process of utilizing computers, databases, multimedia equipment, networks, video equipment and web technologies to electronically collect, classify, copy, compress, scan, store and transform conventional library information resources. Library digitization is different from a digital library as it focuses on the process of making diverse library information resources electronically available, while a digital library is a platform for accessing, collecting, managing, searching and storing distributed digitized information resources over the Internet and World Wide Web (www). (Li). Thus, to provide access to these digitalized library collections, academic libraries need to set up and implement digital library projects which provide digitalized resources, network access and distribution management via network technology.

Digital Library

A digital library is an assemblage of digital computing, storage, and communication machinery together with the content and software needed to reproduce, emulate and extend the services provided by conventional libraries based on paper and other material means of collecting, cataloguing, finding and disseminating information. A full service digital library must accomplish all essential services of traditional libraries and also exploit the well-known advantage of digital storage, searching and communication. (Chowdhury and Chowdhury, 2003). It provides access to part of or all its collections, such as plain texts, images, graphs, audio/video materials and other library items that have been electronically converted, via the Internet and www.

Instant Messaging Reference Service

It is one of the real-time electronic consulting and reference offered by academic libraries via specific software running on the Internet platform. (Li). It is virtually instantaneous communication between two or more people using textual format, providing "real time reference" services, where patrons can synchronously communicate with librarians much as they would in a face to face reference context. The software often allow co-browsing, file sharing, screen capturing and data sharing and mining of previous transcripts. Libraries are already offering live reference service using 24×7×365 in a collaborative fashion.

Information Commons

An Information commons is an innovative and evolving collaborative academic library service model built on a variety of networked interactive academic learning platforms. The primary function of an academic library information commons is to integrate existing information resources, services, instructions and other public service programmes in the library into one consistent dynamic, interactive and scalable student centered interactive academic learning environment.

It is also called a learning commons, which serves an integrated one stop information gateway for users of the library.

Wiki

A Wiki is a website that uses Wiki software, allowing the easy creation and editing of any number of interlinked web pages, using a simplified markup language. Wikis are often used to create collaborative websites, to power community websites and for note making http://en.wikipedia.org/wiki/Wiki.For example, the collaborative encyclopedia, Wikipedia is one of the best-known Wikis that has broken down the golden rules of library science, i.e. content validation and authentication of information. Wikis are also used in business to provide affordable and effective Intranets and for knowledge management. Libraries can use Wikis as a communication tool to enable social interaction among librarians and patrons. Users can share information, ask and answer questions, and librarians can do the same within a Wiki. Moreover, a record of these transactions can be achieved for perpetuity. Transcripts of such question-answer sessions would serve as resources for the library to provide as reference. A Wiki like platform created for the librarians to work collaboratively and concurrently on providing answers to the user enquiries. This allows any staff to tap on the collective wisdom of the communities of subject librarians and provide quality answers to their queries.

Blog

A blog (an abridged form of term web log) is a website, usually maintained by an individual, with regular entries of commentary, descriptions of events, or other material such as graphics or video.(http:// en.wikipedia.org/wiki/Blog). Blogs provide control to an individual or group of individuals for publishing contents or making commentary on it. Technologically, blogs are easier to use, platform-independent and accessible online over the Internet. Blogs are increasingly used by libraries as promotional, alerting and marketing tools; providing a useful method of promoting new services, alerting users to changes and offering advice and support. In library blogs typical posting include information about fresh arrivals, edatabases, news and services rendered can be flashed for wider effects.

Disseminator of Information Literacy

ALA defined information literacy is "recognizing when information is needed, and having the ability to locate, evaluate and use effectively this needed information." (http://en.wikipedia.org/wiki/Information_Literacy). In 2003, ALA evolved this definition and set information literacy standards for student learning, strives that it is essential for higher education institutions, students and staff be provided with opportunities to learn not only how to access information sources but also how to evaluate, manage and use them effectively. Information literary forms the basis for life long learning and enables learners to master content and extent their investigations to become more self-directed, thus assuming greater control over their own learning. This leads information literate individuals to address:

- Assess the extent of information needed;
- Access the desired information effectively and efficiently;
- Use information effectively;
- Evaluate information and its resources critically; and
- Incorporate selected information into their knowledge base. (Nyamboga, 2004).

The library has to provide information literacy services to introduce known and unknown information sources to users by discharging its prime function of service oriented knowledge centre.

Future Visions of Library in E-resources Environments

Visions of Technological Changes

The libraries will utilize multiple media extensively. A visual infrastructure of video-displaying walls, situation room theaters, learning "cafeterias," and dispersed, theme-centered constructions utilizing multi-media "books" and other knowledge-based packages, exhibits, arcades and laboratories ect.

Virtual Conferencing

Anticipates virtual conferencing with access to extensive media storage, providing opportunities for students to explore issues and locales much like journalists learn their way in a new foreign assignment. Bill Kennedy, a university Webmaster, envisions similar uses of technology, but he describes the situation in terms of metaphors. No longer a room with a host, the library of 2012 will be experienced as a virtual reality with a "zoom atlas" to whisk the learner to other places, with time travel to jump back into history or forward into the future, and with enacted dialogue to allow "conversations" with people from other times and places. Not one metaphor, or a few, but a virtual "Metaphor Factory" is the vision Kennedy offers. These two essays share visions of continuous media providing the means to escape existing constraints.

Cybrarians in InfoSpace

"Cybrarians in InfoSpace" is the theme of the winning essay, submitted by a team of library school professors. In their view, learning is the theme of the day, with "cybrarians" heavily engaged with students both individually and in learning clusters. Surprenant and Perry envision librarians as technologists, working with tools that utilize artificial intelligence and multitasking to assist learners in creating individualized information portfolios. Communicating through Virtual Reality helmets and V-mail, and utilizing diagnostic tools to customize resources to individual profiles, cybrarians will provide effective support for problem solving and discovery groups.

Wild Card Libraries

"Wild Card Libraries" is the vivid image offered digital harvesting of information, and knowledge, for purposes of extensive "content building" projects initiated by research libraries and facilitated by Internet 3 technologies will result in malleable, globally linked archives of knowledge and information.

Visions of Change in Library Functions

From Place to Function

"From Place to Function" is the theme offered by Alan Bailin (Baruch College, CUNY) and Ann Grafstein (Hofstra University). What marks the day will be the range of services offered (face-to-face and at a distance), to package electronic documents and resources generated primarily by libraries rather than by commercial publishers. Selective dissemination of information and alert systems linked to a student's course management account and supported by virtual reference will highlight library services in the future. The library will be ubiquitous, and its range of services will dramatically overpower the roles related to the traditional library as "place" with its books and printed materials.

Working the Network

"Working the Network" theme offered by Kelsey Libner, a North Carolina State University Library Fellow. With collaboration as the key, librarians will work closely with other libraries, information technology and computer science departments, instructional designers, and information architecture specialists to service student needs. Customization and personalization are seen as key value-added contributions. Multiple collections, some designed for a specific course, utilizing video clips and various media, will simplify the complex morass of information for the learner. Access to preprints, software, listserv archives, websites and other currently overlooked resources will become commonplace and necessary.

Telling Many Stories

"Telling Many Stories" offers a distinctive vision of providing artifacts and sensations, of managing "living" documents to enhance learning and experience. Publication archives and "experience depositories" will result from enabling students to understand culturally distinctive "ways of knowing" and creating bio-personal and bio-social information and research portfolios, kept current with intelligent agents. (This account, from David

Brier of the University of Hawaii, Manoa, could obviously be considered more technological than functional, but the "story-telling" thrust tilts the balance to categorizing the essay as one dealing with function.).

Visions of Librarians' Roles

A Librarian in the House

This third theme stresses the personal role of the librarian, though the essays emphasized very different things. Beth Posner (City University of New York Graduate Center) asks if there is "A Librarian in the House?" Her conviction is not that librarians will disappear, but rather that they will be out and about, preserving the traditional mission of the library by proactively calling on colleagues and making face-to-face presentations to professors, departments and classes any time such opportunities can be created. Julie Still, another Rutgers University, Camden, librarian suggests that librarians will hold their place simply because they are "So Darned Charming". She insists that "self-service" information finding will not push the profession aside. Drawing on personal experience, she points out that while she could change her own oil, sew her own clothes, and cook every meal, she won't. People will continue to come to librarians because they lack the time and skills to efficiently do the job themselves.

Conclusion

Changes are inevitable thus, ignoring the change leads to failure and acceptance trails to success. The challenges associated with converting of print resources to e-resources and preserving the information resources required for the support of the users in the libraries have never been more complex and demanding than they are today. Plethora of information and variety of modes pose challenges but acceptance and harnessing technological developments by library professionals will help in quenching the quests of information seekers, redefining the nature, role, services and value of libraries and proving its nucleus essence for the social and economic development of the society.

References

1. **Arant, Wendi and Payne, Leila.** "*The common user interface in academic libraries, myth or reality?:Library Hi-tech Journal"; Vol. 19, Issue 1,* 2001: 63-76p.
2. **Breeding, Marshall.** *"Managing Resource Somprehensively: Computers in Libraries"; Vol. 28 Issue 8,* , Sep2008: p28-30, 3p.
3. **Carlson, Scott.** *"The Deserted Library: Chronicle of Higher Education; Vol. 48, Issue 12,* 2001: 35p.
4. **Chase, Linda and Dygert, Claire.** *"Organizing Web-based resources: Serials Librarian"; vol. 38, Issue 3/4,* 2000: 277p.
5. **Christine Urquhart, et.al.** *"Uptake and use of electronic information services: trends in UK higher education from the JUSTEIS project".* 2003: 277*p.*
6. **Curran, Mary.** *"Organize, Simplify and Realize: Using the SFX Monthly Update Reports to Manage New and Dropped Titles in a Matter of Minutes: Serials Librarian"; Vol 51, Issue 1,* 2006: 27-36, p.
7. **G.F.** *"States Offer At-Home Database Access: American Libraries"; Vol. 30, Issue 2,* 1999: 82p.
8. **Greenberg, Jane.** *"Meta views in the millennium Research and Education: Journal of Internet Cataloging"; Vol. 6, Issue 3,* 2003: 77p.
9. **Harper, Paulina, et.al.** *"The 1st Electronic Resources and Libraries Conference: A Report: Library Hi-Tech News"; Vol. 23 Issue 5,* 2006: 12-22p.
10. **Hawkins, Les.** *"Refinement of Cataloging Tools: Serials Review"; vol. 26 issue 4,* 2000: 37p.
11. **Helmetsie, Carolyn L. and Hopkins, Randall M.** *"Managing Multiple Media and Extraordinar expectations: Serials Librarian".* 2000: 225p.
12. **Klobas, Jane E.** *"Beyond information quality: fitness for purpose and electronic information resource use: Journal of Information Science"; Vol. 21,* 1995: 60-65p.
13. **Lindley, David.** *"Smart selection, streamlined processes: Library & Information Update"; Vol. 6, Issue 9,* 2007: 35-37p.

14. **Morrison, Peggy.** *"Management Column: E-metrics—Getting There"* 2004: 20-22p.

15. **Mudhol, Sujata H.R. and Mahesh V.** "*Use of Electronic Resources in the College of Fisheries: Annals of Library and Information Studies"*. 2008: 55p.

16. "NEWS IN BRIEF." *Advanced Technology Libraries"; Vol. 34, Issue 4,* 2005: 12-12p.

17. **Sharma, S. K.** *Information Technology and Library Services.* 2007: 155p..

18. **Tennant, Roy.** *"The Library Brand: Library Journal"; Vol. 1, Issue 1,* 2006: 38-38p.

19. **Thapa, Neelam.** *Slices of Library Automation, Information Technology.* 152, p. 2007.

20. **Wells, Andrew.** *"Looking at Licenses"*. 2007: 30-40p.

21. **Waldman, Micaela.** "*Information Research*", Vol.8 No. 2003. Freshmen's use of library electronic resources and self-efficacy.

22. **Corpenter, Todd.** "*Agaomst the Grain*" (Dec 2007- jan 2008). Electronic Resource Challenges and Opportunities. http://www.against-the-grain.com

23. **Angrosh, M.A.** "*Electronic and Print Journals : A Value Chain Analysis*". Annuals of Library and Information Studies Vol. 52, 2; 2005; p56-64.

24. **Singh, R.K. Joteen, Madhuri, Th. And Chaudhury, Arup Ray.** "*Use of Internet Based E-resources at Manipur Universiyt : A Survey*". Annuals of Library and Information Studies Vol. 56, March 2009; p52-67.

25. **Constantine, M. and Nyamboga, T.D.** (2003). "*Annals of Library and Information Studies 50*", Development of libraries and information centres in the electronic age.

26. **Bell, A.** "*The Internet*": a new opportunity for information specialist. Information outlook. 1, 9; 1997; 33-36.

27. **Mounissamy, P and Kaliammal, A.** (2006). "*Promoting Effective Use of Electronic Resources Using Library*" websites by IITs and NITs : a comparative study. IASLIC Bulletin vol.51 (4). P213-220.

28. **Natarajan, K and Suresh, B.** "*Use and User Perception of Electronic Resources in Annamalai University: a Case Study*". Annuals of Library and Inforamtion Studies Vol57. March 2010, p59-64.

29. **Parinsen, Jola G.B.** (2001). "*A Challenging Future Awaits Libraries able to Change*". D-Lib Magazine, 7.

30. **Uma, K.** "*Migrating to the Electronic Learning Environment: Prospects for LIS Distance Learners in India*". (PDF file accessed through Internet)

31. **Manish, Kumar.** *ICAL* (2009). "*Vision and Roles of the Future Academic Libraries*".

32. http://alpha.fdu.edu/~marcum/visions_silverstone.doc

33. http://alpha.fdu.edu/~marcum/visions_silverstone1.pdf

34. **Thamaraiselvi, G.** Vision and the Changing Role of the Future Academic Library Professional in the E-learning Environment: Challenges and Issue in ICAL 2009–Vision and Roles of the Future Academic Libraries.

31

Ethical Considerations and Perspectives for Library Science Professionals

Ms. Ruchika Krishna &
Manoj Srivas

Abstract

Three major issues are addressed in this discussion: (1) the role, status, and compensation of such non-professionals as library clerks or technicians vis-a-vis professionals, the organization, and the public, particularly in their claims for, or realization of, professional status; (2) the role, authority, status, and compensation of non-librarian professionals appointed as directors or supervisors; and (3) the relationship of professional librarians to other professionals on the library or information center staff. After characterizing the nature of a library professional, the actual and theoretical criteria for such a designation are discussed. Non-professional librarians may argue and strive for such status, but there are many that should be considered. There are many stakeholders, a variety of ethical principles (e.g., such principles as seeking justice or fairness or preserving professional or organizational trust), a variety of ethical obligations (e.g., obligations to the self, the organization, or society), diverse loyalties (e.g., to the profession or the organization), and varying circumstances and conditions, each of which must be brought into ethical deliberation. For each of the major issues, this article delineates the perspectives, values, obligations, and priorities stakeholders bring. In such a manner, the complexity and diversity of factors will be made clearer so that resolution, if it can occur in a particular case, can serve the best ideals or seek a working consensus

Introduction

Professionalization of planning and the credentialing processes that will accompany it add importance to a review of the role of professional ethics in planning. Existing ethical standards are often inherently contradictory, guild oriented, and inconsistent with the public image the profession attempts to maintain. The more publicly oriented prescriptions are not designed to be enforced. Generally they are a weak guide to ethical conduct for practicing planners.

Planning theories suggest ethical standards going beyond professional prescriptions. A historical or structural approach would further suggest that professionally derived standards will be inherently system maintaining and that efforts to inject more progressive and enforceable guidelines into professional codes are likely to meet major resistance within the profession.

Objectives

1. To study **the heritage and traditions of international librarianship** and begin rethinking these traditions in light of current challenges.
2. To study the historical and philosophical background of **ethical traditions and various strategies for decision-making.**
3. To understand the **social contexts** that have shaped the values and moralities of the successive information ages,

the history of reading, writing, and literacy, and the rise of contemporary information and communications technologies.

4. To become comfortable with the **vocabularies and methodologies** of applied ethics, STS (Science, Technology, and Society), social informatics, and the philosophy of technology.
5. To **practice critical thinking** and various approaches to decision-making in **supportive environments** where many different voices can be heard.
6. To learn to apply the values represented in **professional codes** yet also to think and act beyond the confines of established codes.
7. To appreciate **global dimensions** of ethical, legal, and cultural issues.
8. To learn to use resources that will guide them throughout their **careers** to explore and evaluate ethical challenges, the professional literature, the work of organizations and associations both inside and outside the field of LIS, governmental, and non-governmental policy resources.
9. To become familiar the **research** literature and methodologies of the various academic disciplines contribute to understanding of global information ethics
10. To develop a **professional perspective** to guide them toward personal integrity and social responsibility in the work place and in their participation in larger society.

The Codes

A code of ethics is a list of guiding principles for ethical behavior. For example, a code of ethics tends to contain statements of the form "You shall do X" (e.g., "You shall protect intellectual property rights") or "You shall not do Y" (e.g., "You shall not censor library resources"). Codes of professional ethics for library organizations are mainly intended to guide the behavior of library professionals. However, these codes serve other functions as well. In particular, these codes of professional ethics inform the public about what library professionals are committed to doing.

In a survey of library professionals, Wallace Koehler et al. (2000) found that there are differences in which principles are emphasized, but that there is fairly wide agreement about what the principles are. For example, library professionals should "uphold the principles of intellectual freedom and resist all efforts to censor library resources." Also, they should "protect each library user's right to privacy and confidentiality" and "recognize and respect intellectual property rights."

Codes of professional ethics for library organizations typically address the core issues of information ethics, such as intellectual freedom and intellectual property. But it should also be noted that these codes typically go beyond principles of information ethics. They include principles of professional ethics more generally. For example, they typically discuss such issues as the treatment of employees and professional development. But such issues are common to many professions and are not directly related to information. This paper will focus specifically on the principles of information ethics.

The Theories

Ethical theories can be roughly divided into four main types depending upon whether they appeal to consequences, duties, rights, or virtues. For each of the aforementioned ethical dilemmas, we might profitably ask what a consequence-based theory, a duty based theory, a rights-based theory, or a virtue-based theory would say that we should do. Indeed, answering this question is a useful exercise for students.

a) Consequence-based Theories : According to a consequence-based theory, what distinguishes right actions from wrong actions is that they have better consequences. In order to do the right thing, we should perform actions that have the good *consequences*. Consequence-based theories clearly have quite a bit of intuitive appeal. Also, they can be easily applied to the ethical dilemmas faced by library professionals. The main example of a consequence-based theory is *utilitarianism*. According to utilitarianism, goodness is measured in terms of the amount of happiness in the world. Thus, the right action is the one that maximizes overall happiness. The most influential development of utilitarianism is due to the British philosopher John Stuart Mill (1863). Mill's (1859) influential argument for intellectual freedom, and against censorship, shows how utilitarianism can be applied to information ethics. There are two steps to Mill's argument. First, he argues that we are more likely to acquire true beliefs if there is no censorship. Second, he argues that acquiring true beliefs tends to increase overall happiness. In support of the first step, Mill points out that, since human beings are fallible, we are sure to censor some true information if we censor (even if we try to only censor false information). Furthermore, even if we succeeded in only censoring false information, our true beliefs would quickly become "dead dogma[s], not ... living truth[s]." That is, we would lose the conviction in our beliefs that comes from seeing how they stand up to criticism.

Tony Doyle (2001) has recently argued that Mill's argument actually supports an *absolute* ban on censorship. Doyle (2001, 60) admits that, according to utilitarianism, "if we could be sure that a type of expression was seriously inimical to half the population, that it carried no compensating benefits, and that it could be banned with few repercussions, then we should ban it." However, he still argues that there should be an absolute ban on censorship because *we cannot be sure* which types of expression will have these bad consequences. Don Fallis and Kay Mathiesen (2001) have argued (contra Doyle) that there may be cases where the potential consequences are sufficiently bad and sufficiently likely that censorship is unfortunately the right action. Consequence-based theories can easily be applied to other issues in information ethics as well. For example, Edwin Hettinger (1989, 47-51) has offered a utilitarian argument for respecting intellectual property rights. The basic idea is that, if intellectual property rights are not respected, authors will not be able to recover the costs of producing the intellectual property. As a result, they may not be willing to create (and supply libraries with) more intellectual property, which would clearly be a bad consequence.

***b*) Duty-based Theories :** Consequences are not necessarily all that matters in determining what the right thing to do is. Many ethical theorists think that there are ethical *duties* that human beings must obey regardless of the consequences. For example, we arguably have a duty not to kill innocent people even if doing so would have very good consequences. The most influential duty-based theory was developed by Immanuel Kant (1785).

According to Kant, the basis for right action is the *categorical imperative*, which states that "I should never act except in such a way that I can also will that my maxim should become a universal law." It follows from this, for example, that lying is wrong. If *everybody* lied, then no one would trust anybody else and there would be no point to lying. Thus, the maxim "Lie whenever it is to your advantage" would not work as a universal law. Kant gives other versions of the categorical imperative that actually provide more straightforward guidelines for identifying right actions. For example, a well-known version states that you should "act in such a way that you treat humanity, whether in your own person or in the person of another, always at the same time as an end and never simply as a means." In other words, you should not simply *use* other people in order to achieve your goals. Woodward (1990, 15) has tried to use Kant's categorical imperative to provide a defense of intellectual freedom.

A more recent (and more user-friendly) duty-based theory was developed W. D. Ross (1930). One reason for the greater accessibility of Ross's theory is that (unlike Kant or Mill) he does not try to distinguish between right actions and wrong actions using a single unified principle. Ross instead presents a whole list of duties that are each supposed follow directly from our moral intuition. This list includes a duty to keep our promises, a duty to distribute goods justly (*justice*), a duty to improve the lot of others with respect to virtue, intelligence, and happiness (*beneficence*), and a duty to avoid injury to others. The duties of justice and beneficence are especially important for library professionals. In addition, Ross's list of duties is not intended to be exhaustive. As a result, there may be additional duties (possibly a duty to provide access to information) that are directly relevant to library professionals.

c) Rights-based Theories : Other ethical theorists think that the right thing to do is determined by the *rights* that human beings have.[9] The most influential rights-based theory was developed by John Locke (1689). Such theories are especially congenial to information ethics as discussions of these topics are often framed in terms of rights. The *Library Bill of Rights* (ALA 1996) is a notable example.

We have some rights merely by virtue of being human. For example, the "inalienable rights" that Thomas Jefferson appeals to in the Declaration of Independence are of this sort. Such *natural rights* have many potential applications to information ethics. For example, it has been suggested that it is in the nature of human beings to think for themselves and that this fact implies that we have certain rights. In particular, Woodward (1990, 15-16) claims that this fact establishes that we have a natural right to unrestricted access to information. If information were generally withheld from us, our ability to think for ourselves about what we should do would be seriously impeded. In other words, restrictions on access to information would conflict with our nature. Along similar lines, it has been argued that we have a natural right to privacy (see, e.g., Schoeman 1984). The basic idea is that we are not really able to think for ourselves if we are worried that our choices (e.g., about what to read) are being observed. In addition to our natural rights, we also have rights that arise from our participation in a society. John Rawls (1971) has developed a very influential ethical theory that focuses on these sorts of rights. People do not usually make an explicit agreement to participate in a society. Furthermore, even if they did, it is not clear that such an agreement would be *fair*. The people who have more power often take advantage of the people who have less power. As a result, we cannot base an ethical theory on an actual agreement that people have made. Instead, Rawls bases his theory on the idea of a *hypothetical* (but fair) agreement.

In order to determine what a fair agreement would look like, Rawls asks us to perform a thought experiment. We imagine that the parties to the agreement are behind a *veil of ignorance*. That is, we imagine that they do not know anything about their particular position in society. For example, they do not know what they have (e.g., how wealthy they are), what their abilities are (e.g., how smart they are), or what goals they have in life. In other words, they do not know anything about themselves that might bias their decisions about what policies to adopt.There are a number of ways in which Rawls's theory can applied to issues in information ethics. For example, Martin Frické et al. (2000, 482) have argued that Rawls's theory supports public funding of, and equitable access to, library services. Since a person behind the veil of ignorance does not know what her position in society is or what goals she actually has (i.e., what her "conception of the good" is), she cannot simply adopt policies that support the goals that she actually has. However, she can be fairly certain that she will need access to information whatever her specific goals happen to be. In other words, as Jeroen van den Hoven (1995) points out, information is what Rawls's would call a "primary good." Thus, she will want to make sure that access to information is provided to *all* members of society. Even if she is actually rich, she would agree to the public funding of library services because she has to allow for the possibility that she is poor and really needs those services. Rawls intends his theory to be used to evaluate large-scale social policies (such as whether libraries should be publicly-funded). But his theory can also be used to evaluate small-scale social policies (such as how such libraries should be run). For example, Wendell Johnson (1994) has used Rawls' theory to evaluate reference policies. Also, Rawls's theory can be used to defend the intellectual freedom policies that most libraries adhere to. Since a person behind the veil of ignorance does not know what specific information she will need, she will want to make sure that access to information on a wide range of topics and from a wide range of perspectives is provided. Rights-based theories can be applied to other issues in information ethics such as intellectual property. Adam Moore (2001), for example, has offered a Lockean defense of intellectual property rights. That is, he applies Locke's influential theory of property (which basically says that we have a natural right to the "fruits of our labor") to the special case of intellectual property. However, Thomas Jefferson (1813, 630) has claimed there is an important disanalogy between intellectual property and other sorts of property. He writes that "he who receives an idea from me, receives instruction himself without lessening mine; as he who lights his taper at mine, receives light without darkening me." Hettinger (1989, 36-45) offers additional objections to a Lockean defense of intellectual property rights. Roger McCain (1988, 270) also points out that intellectual property rights can come into conflict with other property rights. For example, my right that my book not be copied without my permission conflicts with your right to use your expensive photocopier as you see fit (compare the dilemma for the school librarian described above). It is not immediately obvious which right takes precedence in such a conflict.

d) Virtue-based Theories : Finally, a few ethical theorists think that the right thing to do is determined by the *virtues* that human beings ought to have. According to virtue-based theories, the right thing to do is what a virtuous person would do in the same circumstances. The most influential virtue-based theory was developed by Aristotle (350BC). The Aristotelian virtues include things like courage, temperance, friendliness, and generosity.

Virtue-based theories have probably been the least discussed of the four types of theories. But they have been gaining in popularity in recent years. Philippa Foot (1978) and Alasdair MacIntyre (1981), for example, offer contemporary virtue-based theories. And virtues are clearly applicable to issues in information ethics. For example, library professionals often need to courage to stand up for the principles of information ethics in the face of resistance. Also, as Ashley McDowell (2002, 56) points out, friendliness certainly makes it more likely that library professionals will succeed in their mission of providing people with access to information.

Conclusion

Library and information professionals play an extremely important role in society. And it is critical that they carry out their mission in an *ethical* manner. Toward this end, many organizations of information professionals have adopted *codes of professional ethics*. The various ethical theories that have been proposed by philosophers. We will then apply these theories to the ethical dilemmas that most commonly confront library and information professionals. In particular, we will focus on issues such as intellectual freedom, equitable access to information, privacy, and intellectual property. In addition, we will look at how advances in information technology have created new ethical dilemmas.

References

1. **Alfino, M., and Pierce, L.,** (1997), "*Information ethics for librarians*", McFarland, Jefferson, North Carolina.
2. American Association of Law Libraries, (1999), Ethical principles, available at: http://www.aallnet.org/about/policy_ethics.asp
3. American Library Association, (1995), Ccde of ethics, available at: http://www.ala.org/ala/oif/statementspols/codeofethics/codeethics.htm
4. American Library Association, (1996), Library bill of rights, available at: http://www.ala.org/ala/oif/statementspols/statementsif/librarybillrights.htm
5. American Library Association, (2002), Intellectual freedom and censorship Q & A, available at: http://www.ala.org/ala/oif/basics/intellectual.htm
6. American Society for Information Science and Technology, (1992), Professional Aristotle, (350BC/2000), Nichomachean ethics, Cambridge, Cambridge.

32

Trends of Electronic Tools : A Comparative Analysis of Search Engines

Rajesh Kumar Gupta &
Naveen Kumar Srivastava

Abstract

The study has been evaluated the retrieval of information efficiency of five common Internet search engines, by using sample queries under which four simple queries even last four belongs to Boolean queries in the subject of LIS. For each search engines the total number of sites, selected sites (Ten Number), relevant sites and irrelevant site are recorded and presented in the tables. "Google" and "Ask.com" has been found the best search engines for retrieval of relevant information in "simple" and "Boolean" queries respectively in broader sense.

Keywords: *Information Retrieval; Internet; Search Engine.*

Introduction

A search engine is used to locate the information on the Internet. Although each search engine shows differences in the cataloging materials, but they all work basically in the same manner. The user enters a keyword or phrase and the search engine displays lots of hits found in its database related to the users input. If the user does not find sufficient information, they can enter a different keyword or phrase to refine the resemble information to display faster and accurate results.

Five search engines Google, Yahoo, Bing, Lycos and Ask.com has been compared and evaluated in term of their search capabilities (e.g. Boolean logic and phrase search) and retrieval performances (i.e. precision) using sample queries drawn from real reference questions in LIS. The Lycos yields a greater number of output but as far as precision is concerned Google, Lycos & Ask.com seem to be somewhat higher than Yahoo & Bing.

Objective of the Study

For a long time the library was the most used information source for information seeker and library catalogue was the effective source for searching information in their library. But in the digital era, more scholars use the Internet for their information need and search engines are the effective resources for searching information through World Wide Web (WWW).

The main goal of this study is to observe the role of best search engine for retrieval of relevant information.

Sample Queries and The Testing Environment

As a sample queries, the following four research topic in LIS were selected for testing various features of each search engine as well as to represent the different levels of complexity of search. In simple search option we have used simple key word and Boolean operators (and, or, not) use only and (+) operator for Boolean search.

During the search of relevant information, only first ten result performances among all results has been observed.

Query Example

The following reference code has been taken for getting the results in several search engines. These are as follows -

1. "Digital Library Issues"
   - "Digital Library" + "Issues"
2. "Open Access Journals"
   - "Open Access" + "Journals"
3. "Semantic Web"
   - "Semantic" + "Web"
4. "Library Building"
   - "Library" + "Building"

Sample Sources, Methodology

Five search engines/directories e.g. Google, Yahoo, Bing, Lycos & Ask.com are selected for this study. Among the total results against a query/search the first ten results has taken for this study. The results which are display in sponsored links have not been taken in the study.

The keyword/ phrases are searched by giving inverted commas first, e.g. "Digital Library Issues" is called simple search while by using implied Boolean (+) with the phrase "Digital Library" + "Issues" is called Boolean search.

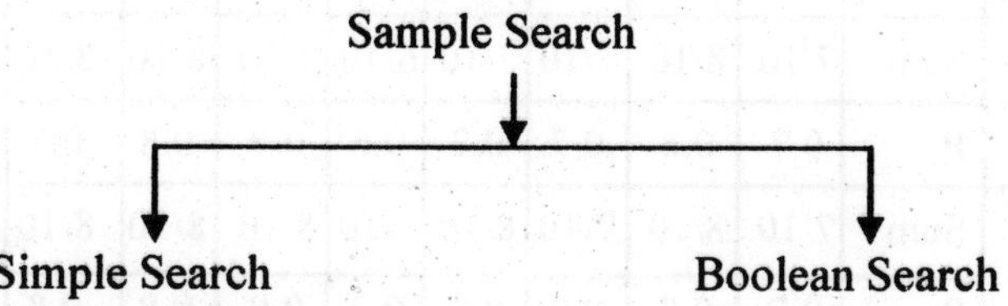

Relevant Sites and Irrelevant Sites

Among the total results against a query/ search the first ten results has been taken for this study and choose as relevant sites which are shows authentic information from any institution, organisation and other relevant bodies in different fields.

While irrelevant sites related to social networking sites like blogs, Wikipedia etc. do not represents the actual statistics/ information authentically.

Observation

Precision of Search Scores : Relevance of retrieved web records are determined on the basis of the upto ten web records downloaded for each query. The investigator has not tried to read the full text web documents by following the links provided because of time constraint and reliability of Web linkage. For each query precision value is computed. The average precisions among twenty searches are also calculated. The mean value of each search engine result is also mentioned in the table no 1.0. In this study the precision score of "Google" and "Ask.com" are higher than Bing, Yahoo & Lycos.

Table 1 : Precision (P) Chart for Five Web Search Engines

Sample Query		#1	#1.1	#2	#2.1	#3	#3.1	#4	#4.1	*Mean* P=N/A
Google	Sum	9/10	3/10	9/10	9/10	8/10	7/10	9/10	9/10	N/A
	P	0.9	0.3	0.9	0.9	0.8	0.7	0.9	0.9	0.79
Yahoo	Sum	8/10	7/10	8/10	8/10	6/10	9/10	7/10	7/10	N/A
	P	0.8	0.7	0.8	0.8	0.6	0.9	0.7	0.7	0.75
Bing	Sum	7/10	8/10	7/10	7/10	6/10	7/10	8/10	8/10	N/A
	P	0.7	0.8	0.7	0.7	0.6	0.7	0.8	0.7	0.71
Lycos	Sum	7/10	8/10	7/10	8/10	7/10	8/10	8/10	8/10	N/A
	P	0.7	0.8	0.7	0.8	0.7	0.8	0.8	0.8	0.76
Ask	Sum	6/10	9/10	6/10	8/10	8/10	8/10	8/10	10/10	N/A
	P	0.6	0.9	0.6	0.8	0.8	0.8	0.8	1.0	0.79
Mean		0.74	0.70	0.74	0.80	0.70	0.78	0.80	0.82	

N/A = Number of Approx; # = Search query.

Table 2 : Total Number of Result (in lac)

Search	*Phase 1*		*Phase 2*		*Phase 3*		*Phase 4*		*Sum*		*Simple+*
Engines	**#1**	**#1.1**	**#2**	**#2.1**	**#3**	**#3.1**	**#4**	**#4.1**	***Simple***	***Boolean***	***Boolean***
Google	1.21	189	14.1	92.9	57.4	119	18.6	753	91	1154	1245
Yahoo	2890	101	28.3	194	193	405	37.9	1370	3149	2070	5219
Bing	2890	2430	105	440	13.4	72.5	2890	1360	5898	4303	10201
LyCos	2890	2200	584	439	72.5	71	2890	1370	6437	4080	10517
Ask	4.59	5.46	44.69	44.69	9.83	42.4	133	352	192	445	637
Total									15768	12051	27818

Table 3 : Comparison of First Ten Results of Relevant Search

SN.	*Search Engines*	*Phase -1*		*Phase -2*		*Phase -3*		*Phase -4*		*sum*
		Simple	*Boolean*	*Simple*	*Boolean*	*Simple*	*Boolean*	*Simple*	*Boolean*	
1	Google	6	3	9	9	8	7	9	9	60
2	Yahoo	6	7	8	8	6	9	7	7	58
3	Bing	3	8	7	7	6	7	8	8	54
4	LyCos	6	8	7	8	7	8	8	8	60
5	Ask	3	9	6	8	8	8	8	10	60

Total number of Result (in lac)

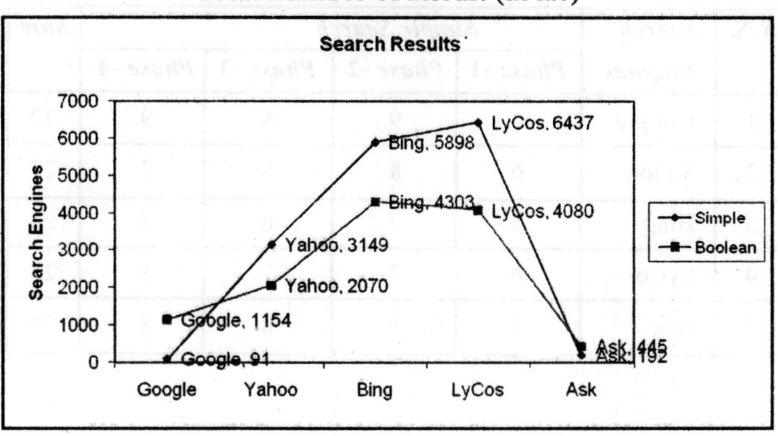

Total number of Relevant Search

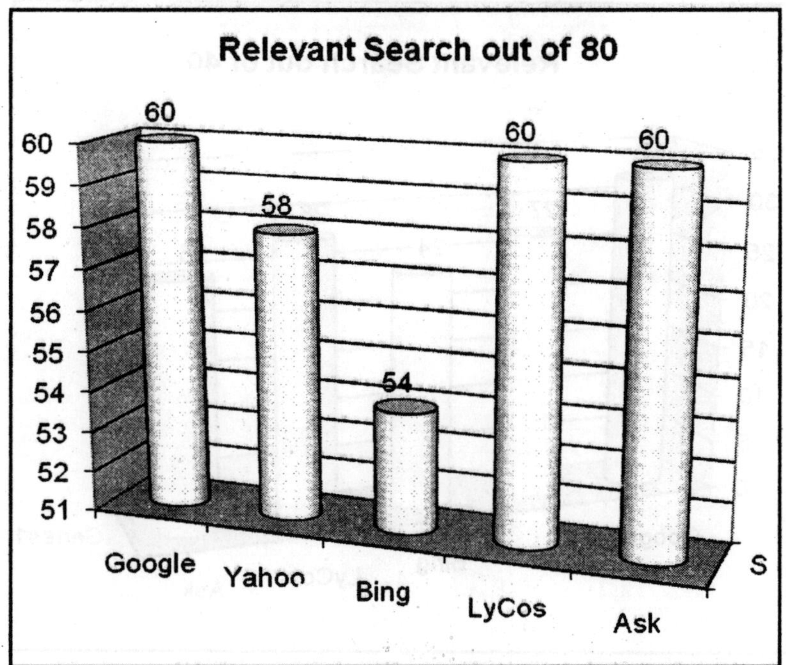

Table 4 : Simple Relevant Search (in first ten results)

S.N.	*Search Engines*	*Simple Search*				*Sum*
		Phase -1	*Phase -2*	*Phase -3*	*Phase -4*	
1	Google	6	9	8	9	32
2	Yahoo	6	8	6	7	27
3	Bing	3	7	6	8	24
4	LyCos	6	7	7	8	28
5	Ask	3	6	8	8	25

Simple Relevant Search

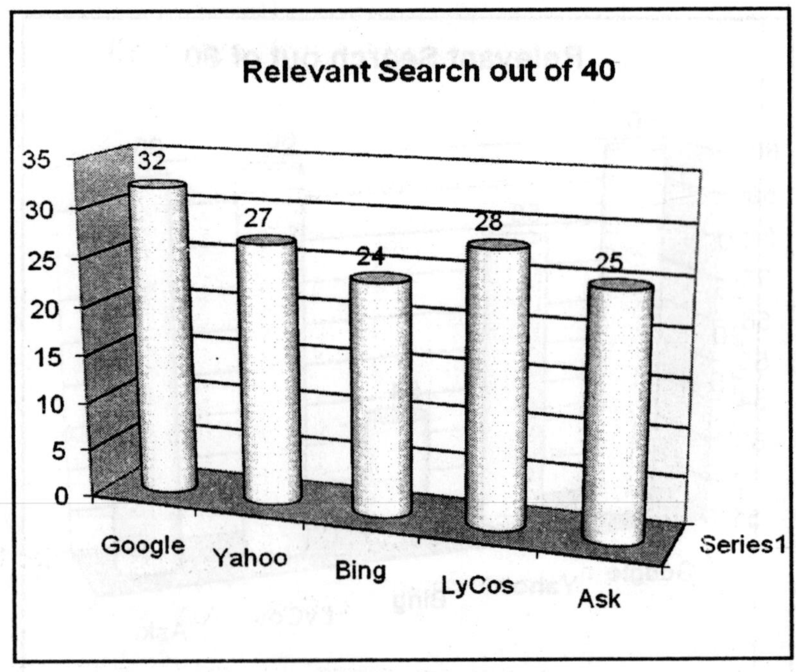

Table 5 : Boolean Relevant Search (in first ten results)

S.N.	*Search Engines*	*Boolean Search*				*Sum*
		Phase -1	*Phase -2*	*Phase -3*	*Phase -4*	
1	Google	3	9	7	9	28
2	Yahoo	7	8	9	7	31
3	Bing	8	7	7	8	30
4	LyCos	8	8	8	8	32
5	Ask	9	8	8	10	35

Boolean Relevant Search

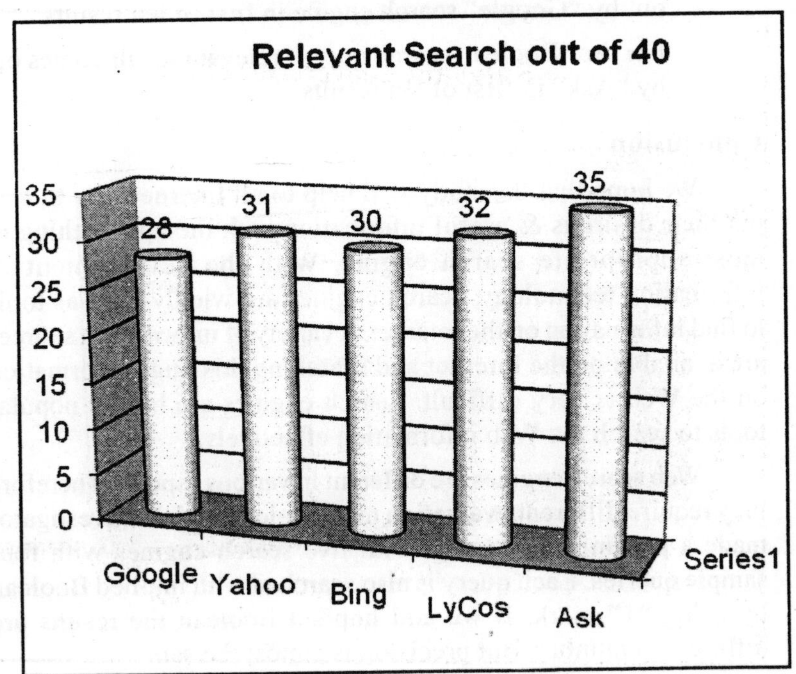

Findings

1. Under sample query we have find the "Lycos" retrieved more search results than any other search engine and as far as precision yielding is most high in case of "Google" and "Ask" while "Lycos" also gives the most precision yielding unexpectedly during the observation.
2. Amongst all search queries "Ask" retrieved minimum search results than any other search engine and as far as precision is same as in "Google" and "Lycos".
3. We also find that for the "Simple search query" maximum number of search result retrieved by "Lycos" while for the "Boolean search query" "Bing" was so higher.
4. For the "Simple search query" minimum number of search result retrieved by "Google" while in "Boolean search query" by "Ask".
5. For the "Simple search query" most relevant result comes out by "Google" search engine in first of ten results.
6. For the "Boolean search query" relevant result comes out by "Ask" in first of ten results.

Conclusion

We hope that the study will help to all Internet user to find out their desirous & useful information with the approaching of most appropriate search engine. With the development of information technology, search engines are widely used as tools to find information on the Internet. A variety of information sources are available on the Internet and searching this huge information on the Web is very difficult. Search engines are highly popular tools to search the Web information effectively.

Web search engines are different in various aspects. Therefore they require different evaluation methodology and the investigator made a preliminary attempt with five search engines with four sample queries. Each query is also searched with implied Boolean by using "+" mark. If we add implied Boolean the results are different in number. But precision is almost the same.

References

1. Balasubramanian, S. and K. Raju (2008). Is Google Enough? A comparative Study of Search Engines. In Proceedings of the NACLIN 2008 Annual Convention, November, 7, 113-30.
2. Biradar, B.S. and Sampath Kumar (2005). WWW Searching Tools: An Evaluation of Features of Selected Searching Tools. In Proceedings of the IASLIC 2005 Annual Conference, February, 33, 355-63.
3. Chu, H., and Rosenthal, M. (1996). Search Engines for the World Wide Web: A comparative Study and Evaluation Methodology. In Proceedings of the ASIS 1996 Annual Conference, October, 33, 127-35.
4. Leighton, H. (June 25, 1995). Performance of Four www Index Services, Lycos, Infoseek, Webcrawler and www warm.
5. http://www.asis.org/ annual.96 /Electronic proceedings/chu.html.
6. http://www.ask.com
7. http://www.bing.com
8. http://www.google.com
9. http://www.lycos.com
10. http://search.yahoo.com

33

Digital Information Service and Challenges

Mrs. Vineeta jain, Mr. Saurabh Nagariya & Prabhat Kumar Pandey

Abstract

Digital library is an emerging trend in the present information technology era. Different services provided with the help of digital library and their issues are mentioned. Digital information service, though looks easy to be provided, poses its own challenges in terms of metadata, digital collection development, copyright, preservation, cost, technology obsolescence, etc., are grouped into different levels like administration, technology and for the users, which are being discussed in the paper. It even highlights how the information professionals can meet the challenges.

Keywords: *Digital library; Library services; Digital information services; Information services; Challenging aspects.*

Introduction

"Right information to the right user at the right time" has been the motto of information professionals. Advancement of Information Technology and emerging trends in Internet has changed the concept of library where the print and paper media are the main components to the new system called "Digital Library". The functions of a digital library are mainly depending upon the computers, communications facilities including equipment used, knowledge and skills of information professionals in respect to handling modern technology. Digital libraries facilitate time and place independent services for users especially if active learning styles become more common place. The digital libraries have several features in addition to computerization of traditional and routine work. These advancements have enabled information professionals to provide the quick and accurate digital information service to the client.

What is Digital Information Services?

Internet, intranets and digital libraries have influenced a lot in providing effective digital information services. Digital information service may be defined as "a service provided to the user with integrate access to different manifestations of information sources which are local, remote, owned, licensed, free, commercial, digital, bibliographic, full text, link, multimedia, data despite of time and distance".

Traditional Information Services Vs Digital Information Service

Traditional Information Services	*Digital Information Services*
Provides services based on print materials	Provides services based on digital and digitized materials
Time and distance bound Does not depend on time and distance	Duplication is more
Duplication is minimized	Personalization is difficult to achieve
Personalization can be achieved	Usage monitoring is difficult to achieve
Usage monitoring is easy to achieve	Information auditing and user needs analysis for identifying Core content and usage preferences
	Resource discovery is fast
	Technology dependent

Types of Digital Information Services

Digital Library & Information Center can provide following Services to its Users :-

- Catalogue Databases,
- Current Awareness Bulletins,
- Externally Purchased Databases,
- CD-ROM Databases,
- Remote Information Services,
- Internally Published Newsletters, Reports & Journals,
- Internet Information Sources Mirroring & Cataloguing,
- E-mail,
- Bulletin Board Service,
- Netnews system,
- File Transfer Protocol (FTP),
- Remote Login (TELNET),
- Audio and Video Communication,
- Electronic Table of Contents,
- Electronic Document Delivery Service,
- Electronic Theses and Dissertations,
- Reference Service,
- Electronic Publishing,
- Discussion groups and forums.

Challenges of Digital Information Services

The Digital Libraries concept is appealing because of its inherent design combines end-user needs with technology that has the ability to handle vast amount of complex data. Perhaps most importantly, the Digital Libraries concept has longevity because it provides services for a society that is identified as being in the process of profound social change including movement towards a knowledge society and the appearance of a weightless economy.

Creating "effective" digital libraries and providing digital information services poses serious challenges for existing and future technologies. The integration of digital media into traditional collections is not straightforward, like previous new media (e.g., video audio tapes), because of the unique nature of digital information-it is less fixed, easily copied, and remotely accessible by multiple users simultaneously. Traditional library processes such as collection development and reference, though forming a potential basis for "digital information service", will have to be revised and enhanced to accommodate these differences.

Digital information services provided by digital libraries, demonstrate a potential for use that extend far beyond being a showcase for technological marvels. As such, digital information services can be viewed as extensions and augmentations of traditional information services by extending the resources that can be offered to the users who can seek and express information. Clearly, Digital information services are becoming an important element in the information cycle and it seems they should also be considered a dynamic, and essential component of library and information science professionals.

Based on the above characteristics, digital information services have to face many challenges. These challenges can be categorized into following 3 types:

A. Administrative point of view i.e., Economics

B. Technology point of view i.e., Efficiency

C. User point of view i.e., Effectiveness

A. Challenges from Administrative Point of View

Setting up a digital library and providing digital information services is not an easy task for the librarian. Librarians have to face various problems like resource crunch, non-availability of required infrastructure, lack of trained human resources, lack of self-exposure, lack of motivation and disinterest of staff etc.

In 1995, Nicolas Negroponte wrote that "being digital" required a transformation from atoms to bits. He used terms such as: multimedia, asynchronous, on demand, open systems, interactive, infinite, multimode, decentralized, harmonized, and mobile to describe the essence of what it would take to become digital. For libraries to achieve this state, they must evolve from library buildings with physical collections and limited access and hours to information services with abundant resources and unlimited availability. Important starts are being made. But formidable problems of access, of delivery vehicles, of developing new forms of navigational aids to locate, create, archive, and preserve content must be overcome. The library community cannot wait for all problems to be solved. We must create new works-multimedia works-that graphically bring alive our collections to audiences both new and old. Our digital libraries will be content-based libraries. Content distinguishes them from other information providers. We will need to re-structure around content rather than function-based activities. Digital collections will require the expertise of systems, cataloging, preservation, selection and service staff they cut across all traditional lines of service and organizational structure.

Metadata: Creating and using new surrogates for non-textual objects is an active area of digital library research. All digital libraries must cope with making metadata available to users. In some cases, only the metadata available are made available digitally. In such cases, users search through pointers and must acquire the primary information physically or through a different system. Huge challenges remain in creating surrogate for digital content. Metadata plays an important role in providing digital information service.

Digital Collection Development: To provide any digital information service one should have sophisticated information sources. As we already know, along with digital content we have legacy content also. These legacy collections should be converted to digitized form to provide information service. An obvious obstacle to digitization is that it is very expensive.

Preservation: To provide continuous digital information service, one should not only have current information sources, but should also have the historical information sources. Preserving these collections is an inherent challenge. To preserve digital information, information on digital hardware and software configuration will require regular "refreshing" or migration to more current systems.

Copyright: Copyright issues are increasing unease among the members of library communities in providing digital collections and services. The problem for libraries is that, unlike private businesses or publishers that own their information, libraries are, for the most part, simply caretakers of information-they don't own the copyright of the material they hold. It is unlikely that libraries will· ever be able to freely digitize and provide access to the copyrighted materials in their collections. This is a great challenge, which needs to be changed by library community to provide effective information services.

Cost: Cost is another challenging aspect in providing information service for the library community. Economic models for making the "digital information services" work, in terms of real costs and benefits, have neither been clearly articulated nor established.

B. Challenging Aspects from Technology Point of View

Technical architecture is another issue that underlies any digital information service system. Libraries need to enhance and upgrade current technical architecture to accommodate digital materials.

The architecture includes the following components:

- High-speed local networks and fast connections to the internet,
- Relational databases that support a variety of digital formats,
- Full text search engines to index and provide access to resources,
- A variety of servers, such as Web servers and FTP servers and

- Electronic document management functions that will aid in the overall management of digital resources

Technological Obsolescence: The major risk to digital object is not physical deterioration, but technological obsolescence of the devices (hardware and software) to read them. Digital libraries are mostly dependent on suitable telecommunication link and computer system for proper utilization and information transfer, these libraries depend much on suitable technology and training of end users in handling a variety of retrieval software, search strategy, formulation and cost consideration in the case of online search.

C. Challenging Aspects from User Point of View

Information in digital libraries is electronically stored and accessed. Access to the digital library is, therefore, no bound by space or time. These libraries provides access to information via electronic gateways to remote digital and digitized databases .The users have become more information conscious. They need pinpointed, updated, comprehensive and relevant information and data pertaining to their special interests and academic pursuits. This has totally changed the users behavior regarding information seeking and using pattern. Accordingly librarians need to formulate the pragmatic strategy for meeting the challenge pertaining to the user behavior with the special reference to the implication of new information technology on libraries.

Meeting the Challenges

The challenge of challenges has added new dimension to the roles and responsibilities of the librarians. These days the librarians are not only the custodian of collection but also content analyzer for documentary as well as electronic information resources for satisfying the increased information needs of the users. These change aspects from the librarians to be well versed with the recent advances regarding the implementation of information technology. On the basis of expense and exposure, the following measures are suggested for the librarians, which will provide adequate base to face the challenges more comfortably;

- Continuous self updating;
- Developing a strong professional team in respect libraries;
- Conducting continuous education and training programs for user and staff;
- Offer training on particular processes and on new aspects.

Conclusion

Digital libraries offer new challenges to an emerging breed of Digital Librarians as new opportunities who should combine principal and practice of information management, with rapidly evolving technological developments to create new information products and services. Hence the information professionals have to keep constant watch for newer developments and noticeable changes in the field of their concern. To cope up with the information needs with speed and relative accuracy and reliability, the digital library is the most important and reliable resort, and so knowledge discovery in these types of libraries become a predominant factor.

References

1. "*The Challenge of the Digital Library*", National Library News: May 1996, vol. 28, no. 5 [online] Available: http://www.nlc-bnc.ca/nl-news/1996/may96e/2805e-07.htm
2. "*Challenges to Building an Effective Digital library*" [online] Available:

   http://memory.loc.gov/ammem/dli2/html/cbedl.html
3. **Kuny, Terry and Cleveland, Gary.** "*Digital Libraries: Myths and Challenges*". Paper delivered at the 62nd IFLA General Conference- August 25-31, 1996. [online] Available:

   http://www.ifla.org/IV/ifla62/62-kuny.pdf
4. **Garrett, John.** "*Digital Libraries : The Grand Challenges*", EDUCOM Review, July/August 1993, Volume 28, Number 4, [online] Available:

   http://www.ifla.org/documents/libraries/net/garrett.txt

5. "*Digital Libraries and Information Access and Distribution*" [online] Available: http://www.internet2.edu/html/digital_libraries.html

6. Digital Libraries and Education : Trends and Opportunities / Hans Roes. - In: *D-Lib Magazine*, vol.7, no.7/8, July/August 2001. [online] Available: http://www.dlib.org/dlib/july01/roes/07roes.html

7. **Joseph S. Alen and John R. Garrett**, "*Toward a Copyright Management System for Digital Libraries*". Salem, Mass.: Copyright Clearance Center, 1991, p. 5

8. **Stephen P. Harter.** "*Scholarly Communication and the Digital Library : Problems and Issues*". Journal of Digital Information 1(1) 1997. [Online] Available:
http://jodi.ecs.soton.ac.uk/Articles/Vo1/ioj/Harter

9. **Guha, B.** "*Documentation and Information: Services, Techniques and Systems Calcutta*", The World Press Private, 1983, p.66.